FERTILIZER

TECHNOLOGY & MANAGEMENT

FERTILIZER
TECHNOLOGY & MANAGEMENT

Brahma Mishra
Ex-Dean, College of Agriculture
G. B. Pant University of Agriculture & Technology
Pantnagar

TECHSAR™
NEW DELHI

Published and Printed by
Techsar Pvt. Ltd.
4435-36/7, Ansari Road, Daryaganj
New Delhi–110 002 (India)
E-mail : tech_info@techsarworld.com
Website: www.techsarworld.com

ISBN 978-93-80026-35-0

Reprint 2025

Preface

Fertilizers are an essential component of the inputs required in modern agriculture as they supplement the soil's nutrient supply to provide optimum nutrition to crops. Fertilizers have played a pivotal role in increasing the agricultural production in developed and developing countries. Fertilizers contributed significantly to the Green Revolution that was witnessed in India and several other countries during 1960s. The quantity of fertilizer used in the world has increased spectacularly during the past four decades, and so has the knowledge about production and management of fertilizers. Not only the product pattern but also the production technology has changed remarkably.

Recent environmental concerns have also put the fertilizers in focus. While some people think of environmental pollution due to fertilizer production and application, there is no gainsaying the fact that fertilizers provide a key input for increasing food production. With increasing population in developing countries, the need of using fertilizers to enhance crop production is increasing and the fertilizers will continue to play a dominant role in agricultural development. There has been substantial progress in the development of eco-friendly production technology and methods for optimizing fertilizer use with little or no risk of environmental pollution.

The knowledge about fertilizer production principles and proper management is not available at a single source. This was the primary felt need which inspired me to attempt this book. The book has been prepared for the readers of various streams such as agriculture, horticulture, environmental science and management. The subject matter has been dealt with in a simple and lucid style to provide easy understanding to beginners as well.

It is hoped that the book will be useful to all who have an interest in the production, marketing and use of fertilizers.

Brahma Mishra

Contents

Preface *v*

1. Role of Fertilizers in Agriculture 1
2. Fertilizer Feedstocks and Raw Materials 15
3. Acids used for Fertilizer Production 27
4. Ammonia Production 35
5. Nitrogen Fertilizers 45
6. Phosphate Fertilizers 63
7. Potash Fertilizers 74
8. Complex and Mixed Fertilizers 84
9. Secondary Nutrient Fertilizers 98
10. Micronutrient Fertilizers 105
11. Fertilizer Legislation and Quality Control 116
12. Fertilizer Marketing 126
13. Fertilizer Logistics 137
14. Fertilizer Application 144
15. Long-term Effects of Fertilizer Use 161
16. Fertilizers and Environmental Pollution 171

Index *182*

CHAPTER

1

Role of Fertilizers in Agriculture

HISTORICAL PERSPECTIVE

Use of commercial fertilizers has only a short history compared to the length of time that man is known to have grown crops. It is believed that crop production began some 6,000 to 10,000 years ago. Before introduction of commercial fertilizers, organic manure was the only external source of nutrients for improving soil fertility and enhancing crop production. The art of manuring for crop production was well known in ancient India during Vedic period (3200 BC-1200 AD). Use of cow dung, bones, and green manure is mentioned in the literature of that period. Green manure and wood ash were commonly used in medieval period (1200-1800 AD). In Chinese agriculture organic manures have been used for over 3,000 years.

Organized research into fertilizer technology began in Europe in the early 17th century. Early scientists such as Johann R. Glauber (1604-1668), a German chemist, described the beneficial effects of the addition of saltpeter (potassium nitrate) to soil. He developed the first complete mineral fertilizer, which was a mixture of saltpeter, lime, phosphoric acid, nitrogen, and potash. As scientific chemical theories developed, the chemical elements needs of plants were discovered, which led to improved fertilizer compositions. In Germany, Justus von Liebig (1803-1873) studied the importance of minerals and atmospheric nitrogen to nourish plants, resulting in his famous Law of the Minimum, stated in 1863, which is a simple but logical guide for predicting crop response to fertilization. This law states that *every field contains a maximum of one or more and a minimum of one or more nutrients. With the minimum, be it lime, potash, nitrogen, phosphoric acid, magnesia or any other nutrient, the yield stand in direct relation. It is the factor that governs and controls....yields.*

The chemical fertilizer industry had its beginning in England with a patent issued to Sir John Bennet Lawes in 1842, which outlined a method for producing a form of phosphate (known as superphosphate) that was an effective fertilizer. The synthetic fertilizer industry experienced significant growth after the First World War (1914-1918), when facilities that had produced ammonia and synthetic nitrates for explosives were converted to the production of nitrogen-based fertilizers.

The world's first commercial nitrogen fertilizer was sodium nitrate mined from natural deposits in Chile. It was imported into Europe and North America from about 1830 onwards. During the 1920s sodium nitrate imports from Chile were still a very important source of nitrogen in the United States and Europe but it diminished with the advent of ammonia synthesis in the middle of 20th century. In India, sodium nitrate was imported till 1960 when large-scale production of urea started.

Of the numerous scientific discoveries in the last millennium, the two that have saved millions of lives from hunger and malnutrition are connected with chemical fertilizers. The first was the manufacture of single superphosphate by Lawes and Gilbert in England in 1843 and the second was the synthesis of ammonia in 1909 by the German chemist, Fritz Haber, who was awarded the Nobel Prize for Chemistry in 1918 for his invention. Carl Bosch, an associate of Fritz Haber, subsequently made the breakthrough necessary to bring the ammonia synthesis process to an industrial scale, thus ushering in the modern nitrogen fertilizer industry. In 1931, Carl Bosch became the Nobel Chemistry Laureate for this accomplishment.

Following the privations of World War II (1939-1945), many countries made food security a top priority. In the following years, policies were put in place to encourage farmers to use fertilizers and other modern farming technologies. Fertilizer consumption grew rapidly, largely in parallel with an accelerating expansion of the world population. The consumption of fertilizer nutrients ($N + P_2O_5 + K_2O$) in the world rose from about 14 million tonnes in 1950 to 27.4 million tonnes in 1960, 116.6 million tonnes in 1980, 137.8 million tonnes in 1990, 138.7 million tonnes in 2001-02 and 155.6 million tonnes in 2004-05. This increase in fertilizer consumption was experienced in all the continents but since 1960 the developing countries have raised their fertilizer consumption impressively as evident from the data shown in Table 1.1. This trend is likely to continue.

Table 1.1 Fertilizer use and annual growth (1959–60, 1989–90, and 2020)

Country category	*Fertilizer use (million tonnes nutrients)*			*Annual growth (%)*	
	1959-60	1989-90	2020	1960-1990	1990-2020
Developed	24.7	81.3	86.4	4.4	0.2
Developing	2.7	62.3	121.6	10.5	2.2
World	27.4	143.6	208.5	5.5	1.2

Source: Bumb and Baanante (1996)

The pattern of global fertilizer consumption has changed markedly over the past 40 years. In 1960, the developing countries accounted for just 10 per cent of all fertilizer consumption; today the figure is nearly 60 per cent. Fertilizer use in the developing world has been fuelled by rapid population growth and growing demand for food grains. This is especially true of Asia, where the scope for land expansion is limited. By contrast, the industrialized countries (with the exception of the former Soviet Union) increased their fertilizer consumption only marginally after 1980; population growth was low, most people were adequately fed, and world agricultural exports had stagnated due to economic problems in many importing countries. Asia is now the dominant player, accounting for 50 per cent of world fertilizer consumption, and 86 per cent of developing countries consumption.

China has been by far the largest fertilizer consumer in the world accounting for 29% of world fertilizer consumption with 44.7 million tonnes in 2004-05. India is another developing country of which fertilizer consumption increased from 0.26 million tonnes in 1961 to 18.4 million tonnes (12% of world fertilizer consumption) in 2004-05. Among developed countries, USA has been the largest fertilizer consumer accounting for 13 per cent of world fertilizer consumption with 20.1 million tonnes in 2004-05.

In India, use of chemical fertilizers began about 100 years ago when a single superphosphate plant was established at Ranipet, Tamil Nadu in 1906. Significant boost in fertilizer use came only after launching the First Five Year Plan in 1951 and subsequently by the Green Revolution (1967-68) which transformed Indian agriculture to bring about much wanted self-reliance in food security.

It has been estimated that about 40 per cent of the protein consumed by humans depends on the Haber-Bosch process of ammonia synthesis. More than 99 per cent of all nitrogen fertilizers are derived from ammonia, which is both an intermediate and a final product. The vast majority of the energy consumed by the fertilizer industry is used for ammonia synthesis (about 94%).

Why Do We Need Fertilizers?

Agricultural production basically depends on the plant's ability to convert solar energy into chemical energy, i.e., food, feed or fibre. This is achieved by plants through photosynthesis with the help of chlorophyll. However, for accomplishing this vital role plants need adequate supply of 13 mineral elements, besides, carbon, hydrogen and oxygen, which are known as essential plant nutrients. These mineral nutrients are divided into two groups: macronutrients and micronutrients. *Macronutrients* are the nutrients that are required in relatively large amounts. They can be broken into two sub-groups: primary and secondary nutrients. The *primary nutrients* or *major nutrients* are nitrogen (N), phosphorus (P), and potassium (K). These major nutrients are usually lacking from the soil because plants use them in large amounts for their growth and survival. The *secondary nutrients* are calcium (Ca), magnesium (Mg), and sulphur (S). There are usually enough of these nutrients in the soil, so fertilization is not always needed. *Micronutrients* are those elements essential for plant growth which are needed in only very small (micro) quantities. These elements are sometimes called minor elements or trace elements.The micronutrients are boron (B), copper (Cu), iron (Fe), manganese (Mn), molybdenum (Mo), zinc (Zn) and chlorine (Cl).

Fertilizers are the materials that supply these essential plant nutrients, mostly nitrogen, phosphorus and potassium, which are removed by crop plants in the largest quantities. Other plant nutrients are required in relatively small quantities.

The mineral nutrient requirement of different crops vary considerably mainly due to yield potential and chemical composition. For example, wheat crop removes more nutrients per tonne of grain produced than the rice does due to differences in the biomass and grain composition. From a study at the International Rice Research Institute, Philippines it was reported that 10–31 kg N, 1–5 kg P and 8–35 kg K were removed from the soil per tonne of rice grain produced. Similarly, for each tonne of wheat grain produced the crop removes 25 kg N, 3.9 kg P and 27 kg K and relatively small amounts of secondary and micronutrients from the soil (Table 1.2). When crop yields of 5–7 tonnes grain ha^{-1} are taken in a single season as with the high yielding varieties of crops, it is not possible for most soils to supply the required amounts of plant nutrients and that is why fertilizers are needed to supplement the soil nutrient supply. Continuous removal of plant nutrients from soil over a long period leads to depletion of soil fertility, which shows up in crop yield decline and lowered factor productivity.

Thus, in a modern crop production system, the plants utilize nutrients from two sources: the indigenous soil supply and fertilizers and manures applied to supplement the soil supply. Leguminous plants like peas, beans, soybean, etc. do not require nitrogen fertilizers as they are usually capable of fixing atmospheric nitrogen to meet their needs after 4-5 weeks since emergence. They, of course, need phosphorus, potassium and other nutrients too. Oilseeds and pulses require more sulphur than cereals, whereas potato, sugar cane and tobacco need more potassium than other crops.

Table 1.2 Nutrient removal by rice and wheat

Nutrient element	*Nutrient removal, kg per tonne of grain*		*Nutrient removal, kg/ha*	
	Rice	Wheat	Rice[1]	Wheat[2]
Nitrogen	16.2	25	113	125
Phosphorus	2.8	3.9	19.6	19.5
Potassium	17	27	119	135
Calcium	4.4	5.3	31	26.5
Magnesium	3.7	4.7	26	23.5
Sulphur	1.7	4.7	12	23.5
Iron	0.24	0.40	1.7	2.0
Manganese	0.65	0.10	4.5	0.5
Zinc	0.04	0.06	0.28	0.3
Copper	0.009	0.02	0.06	0.10
Boron	0.015	0.05	0.09	0.25
Chlorine	3.4	–	24	–

1 For rough rice yield of 7 t/ha (De Datta, 1981)

2 For wheat grain yield of 5 t/ha

Mineral fertilizers contain plant food elements in readily available form, which are essential for plant growth and crop production. Fertilizers are used in order to:

- supplement the natural soil nutrient supply in order to satisfy the demand of crops with a high yield potential and produce economically viable yields,
- compensate for the nutrients lost by the removal of plant products or by leaching or gaseous loss,
- improve unfavourable or to maintain good soil conditions for cropping.

Famine Myths and Food Security

Thomas R. Malthus in his essay published in 1798 argued that unchecked human population would increase geometrically, while the available food supply could only increase arithmetically. He warned that 'the period when the number of men surpasses their means of subsistence has long since arrived'. Two centuries ago when Malthus wrote his essay, the global population was less than a billion. Now the population exceeds 6 billion and in spite of a sixfold increase in human population since 1798 (Fig. 1.1), there is enough food in the market today for all who have the requisite purchasing power.

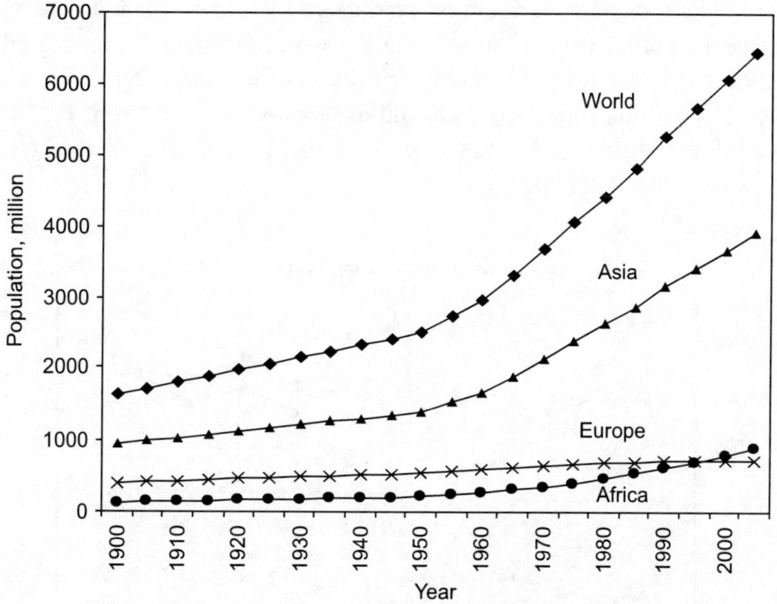

Fig. 1.1 Population growth in Asia, Africa, Europe and world during 20th century

In 1948, Albert Brandt and James Payne published an article on "the world population crisis", which emphasized that "even allowing for spectacular advances in agriculture and industry, the earth simply could not support 4.4 billion people in the year 2048. There would be the constant threat of famine, pandemic disease and unthinkably vicious wars for survival". In the book "Famine—1975" published in 1967 by William and Paul Paddock, it was claimed that the food situation in countries such as Haiti, Egypt and India was so critical that they could not be saved. It suggested that a triage system would have to be introduced, allowing the least fit to starve so that the more robust could be saved. Paul Ehrlich began his 1968 book "The Population Bomb" with the notorious statement, "The battle to feed all of humanity is over" and warned that hundreds of millions of people would starve to death during the 1970s and 1980s.

During the last 50 years, technology and public policy have played a dominant role in averting famines of the kind Malthus anticipated? The last big famine in the Indian subcontinent was the Bengal famine of 1942-43. The spectacular failure of these predictions to eventuate—as exemplified, for instance, by the emergence of India as a significant food exporter—does not seem to have diminished the credibility of those who made them.

Fertilizers and Food Production

The exact contribution of mineral fertilizers to agricultural production is debatable but in any case the millions of field trials which have been carried out throughout the world demonstrate clearly their great influence on crop yields. Fertilizer is the key to securing the food need of the people. No country has been able to increase agricultural productivity without expanding the use of chemical fertilizers. During the period from 1950 to 2005 the world fertilizer use increased 11-fold from 14 million tonnes to 155.6 million tonnes, which resulted in spectacular increase in world food production from 631 million tonnes to 2008 million tonnes. As seen from Fig. 1.2, increase in fertilizer use had direct impact on food production. Similar effect of increased fertilizer use was observed in India and other developing countries.

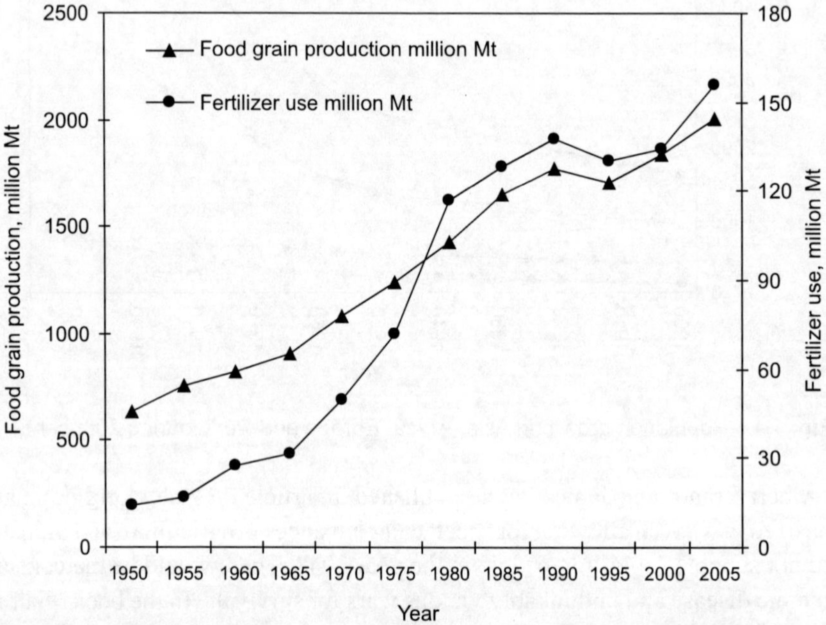

Fig. 1.2 Growth in world fertilizer use and food grain production during 1950 to 2005 (Data source; Earth Policy Institute, & FAI, Fertilizer Statistics)

Stewart and associates (2005) estimated the contribution of commercial fertilizer nutrients to food production from long-term studies representing 362 seasons of crop production. They observed significant variation in crop response to fertilizer inputs, which depends on crop species, soil conditions, climate, geographical location, and other factors. They concluded that at least 30 to 50% of crop yield is attributable to commercial fertilizer nutrient inputs, and that is probably a conservative estimate.

Commercial fertilizers make up the majority of nutrient inputs necessary to sustain current crop yields, with available organic sources, native soil reserves, and biological N fixation supplying the remainder. Using all these inputs efficiently and conjointly is essential in today's agriculture and will be even more important in years to come. To produce the nutritious food supply needed to meet the demands of a growing and more affluent world population, the appropriate and effective use of nutrients supplied from commercial fertilizers is imperative.

Several institutions, among them the FAO, IFPRI, UNDP and the World Bank, have made projections concerning food security. They differ according to the assumptions made, but essentially they are in agreement that the world supply of food will have to keep growing, and growing rapidly. Use of fertilizers, especially in the less developed countries, will be essential, if the objective is to be achieved.

At the turn of 21st century, the world faces many challenges. *First*, food production must be increased to eliminate hunger and malnutrition and ensure food security for all. *Second*, natural resources such as rainforests, wildlife habitats and biodiversity must be maintained. *Third*, soil degradation and nutrient depletion must be halted. *Fourth*, global warming and pollution must be reduced. In confronting all these challenges, increased and environmentally sound use of mineral fertilizers will be essential for enhancing the productivity of cultivable lands, so that increased food could be produced while protecting the environment by sparing forests and marginal lands, replenishing the removed nutrients and sequestering carbon in the biomass production and the soils.

Because of uncertainties associated with future events, it is difficult to make precise predictions. However, one study by IFPRI, Washington estimated that annual fertilizer use in the world should increase to 263 million tonnes ($N + P_2O_5 + K_2O$) to satisfy the food production requirements in 2020. This means that fertilizer use should be nearly doubled in next 15-20 years (141.4 million tonnes in 1999-2000). Given the success of the past (over 5-fold increase in fertilizer use during 1960-1990 period) this may not be a difficult target to achieve but would require a strong commitment from all stakeholders including policy makers.

According to the United Nations, the world's population could increase by 50% during the next 45 years to reach 9 billion inhabitants by 2050 (Table 1.3). With improved lifestyles in today's less favoured countries, global food requirements could more than double. The number of urban inhabitants will, for the first time, exceed the rural population. The area of agricultural land that can be cultivated without degradation of the soil is limited and its productivity must be increased, if intolerable social tensions are to be avoided. This can be achieved only by increase in use of fertilizers particularly in the developing regions.

Table 1.3 Projections of world population (billion)

Year	*More developed region*	*Less developed region*	*Total world*
2010	1.23	5.67	6.91
2020	1.25	6.41	7.67
2030	1.26	7.06	8.34
2040	1.25	7.57	8.82
2050	1.25	7.94	9.19

Source: United Nations

Fertilizer Use and Food Production in India

In India, the consumption of fertilizers witnessed an impressive growth during five decades since 1951-52 when the First Five Year Plan was started. The total nutrient consumption ($N + P_2O_5 + K_2O$)

touched a level of 20.34 million tonnes during 2005-06, the highest so far. The consumption of nitrogen (N) was 12.72 million tonnes, phosphate (P_2O_5) 5.20 million tonnes and potash (K_2O) 2.41 million tonnes. The consumption of total nutrients per hectare of gross cropped area increased from 0.5 kg during 1951-52 to 105.5 kg during 2005-06. Some trends in fertilizer nutrient consumption in India are shown in Tables 1.4 and 1.5.

Table 1.4 Trend of fertilizer production, import and consumption in India (million tonnes)

Year	*Production*		*Import*			*Consumption*			
	N	P_2O_5	N	P_2O_5	K_2O	N	P_2O_5	K_2O	Total
1951-52	0.029	0.01				0.059	0.007	0.00	0.066
1961-62	0.154	0.065	0.142	0.022	–	0.250	0.060	0.028	0.383
1971-72	0.949	0.290	0.481	0.248	0.268	1.798	0.558	0.300	2.657
1981-82	3.143	0.950	1.510	0.452	0.797	4.069	1.322	0.676	6.067
1991-92	7.301	2.562	0.412	1.016	1.326	8.046	3.321	1.361	12.728
2001-02	10.69	3.84	0.164	0.437	1.594	11.31	4.382	1.667	17.360
2005-06	12.72	5.20	1.389	1.145	2.764	12.72	5.20	2.41	20.34

Source: Fertilizer Statistics (2006-07) FAI, New Delhi.

Table 1.5 Trends in intensity of fertilizer consumption in India (kg/ha)

Year	*Fertilizer consumption*			
	N	P_2O_5	K_2O	Total
1951-52	0.44	0.05	0.00	0.49
1961-62	1.60	0.39	0.18	2.17
1971-72	10.88	3.38	1.82	16.08
1981-82	23.02	7.48	3.83	34.33
1991-92	44.15	18.22	7.47	69.84
2001-02	59.44	23.03	8.76	91.23
2005-06	66.00	27.00	12.50	105.50

Source: Fertilizer Statistics (2006-07) FAI, New Delhi.

India is the third largest fertilizer user in the world (China and USA being the first and second), but the consumption is highly concentrated in certain areas with many large areas receiving very little fertilizer. Out of 500 districts (subunits of state), 25 per cent of total fertilizer is consumed in 38 districts, 50 per cent in 102 districts and 75 per cent in 201 districts.

At present the population of India is over one billion and is likely to rise to 1.4 billion by the year 2025. In spite of a rapid rise in population, the food production has kept pace with the needs of the expanding population in recent decades and helped to keep Malthusian fears at bay. This has happened due to *four* factors:

1. Rapid advances in science and technology particularly in plant breeding.
2. Availability of essential inputs such as seeds, fertilizers, pesticides, irrigation and extension services.

3. Public policies in the area of land reform, rural infrastructural development, input and output pricing and marketing.
4. Hard work of farm women and men.

Table 1.6 Progress in food production in India

	1950	*1960*	*1970*	*1980*	*1990*	*2000*	*2005*
Food grain production (million tonnes)	50.8	82.0	108.4	129.6	176.4	201.8	208.6
Food grain import (million tonnes)	4.8	10.4	7.5	0.8	0.3	0.4	1.6
Food grain export (million tonnes)	–	–	–	–	–	2.4	6.9
Buffer stock (million tonnes)	–	2.0	–	15.5	20.8	40.0	21.7
Population (million)	361	439	548	683	846	1000	1104

Table 1.7 Production of food grains and average annual growth (million tonnes)

Year	*Rice*	*Wheat*	*Coarse cereals*	*Pulses*	*Total food grains*	*Compound annual growth (%)*
1950-51	20.58	6.46	15.38	8.41	50.82	–
1960-61	34.58	11.00	23.74	12.70	82.02	3.22
1970-71	42.23	23.83	30.55	11.82	108.42	1.72
1980-81	52.63	36.31	29.02	10.63	129.59	2.08
1990-91	74.29	55.14	32.7	14.26	176.39	3.54
1996-97	81.74	69.35	34.10	14.24	199.44	1.66
2000-01	84.87	68.76	32.22	10.67	195.92	–0.44
2006-07	93.4	75.8	33.9	14.2	217.3	2.09

Source: Dte. of Economic & Stat., Govt. of India

Agriculture accounts for nearly 1/5th of India's gross domestic product (GDP) and more importantly, about 2/3rd of the country's Population is dependent on agriculture and allied activities for their livelihood. Successive Five Year Plans have stressed on self-sufficiency and self-reliance in food grains production and concerted efforts in this direction have resulted in substantial increase in agricultural production and productivity. This is clear from the fact that from a very modest level of 52 million tonnes in 1951-52, food grain production rose to above 206 million tonnes in 1999-2000 (Table 1.6, 1.7). Behind India's success story of not only meeting total requirement of food grains but also having their exportable surplus, the significant role played by chemical fertilizers is well recognized and established beyond any doubt.

Chemical fertilizers have played a vital role in the success of India's green revolution and consequent self-reliance in food grain production. The increase in fertilizer consumption has contributed significantly to sustainable production of food grains in the country. The Government of India has been consistently pursuing policies conducive to increased availability and consumption of fertilizers in the country. The production of nitrogen (N) and phosphorus (P) fertilizer together has increased from mere 3 thousand tonnes in 1950-51 to about 14.7 million tonnes in nutrients terms in 2001-02 and 15.2 million tonnes in 2005-06. Since there are no commercially viable sources of potash (K) in the country, its entire requirement is met through imports.

Fertilizer consumption ($N + P_2O_5 + K_2O$) in India in 1951-52, the first year of the First Five Year Plan was merely 65,800 tonnes (0.5 kg/ha) which increased to 20.34 million tonnes in 2005-06 (105.5 kg/ha). During the same period, total food production increased from 52 million tonnes to 208.6 million tonnes (2005-06). While other factors such as high yielding crop varieties, expansion in area under irrigation and improved crop management practices also contributed to this phenomenal increase in food production in the country, the contribution of fertilizers is estimated to be the largest, about 50% of the total increase. Fertilizers will continue to play a pivotal role in agricultural production in future, because there is little or no scope of expansion in the area under cultivation, hence, more food grains will have to come through increased production per unit area per unit time. Fertilizers are said to have "land sparing action" as they allow more crop production from less land.

The growth of fertilizer consumption in India has not been uniform either in time or in space. There are regional differences, seasonal differences and annual ups and downs. Of particular interest is the impact of economic liberalization policies introduced in 1991-92, which raised the price of fertilizers due to partial withdrawal of govt. subsidy and depressed the rate of growth in fertilizer consumption. However, the average crop yields in different states in India reflect the direct effect of fertilizer use as evident from Figs. 1.3 and 1.4. In Fig. 1.3 the average yield of rice (clean rice) is plotted against the average fertilizer use for the year 2003-04 in 17 rice growing states in India. Similarly, in Fig. 1.4 the average yield of wheat is plotted against the average fertilizer use for the year 2003-04 in 9 wheat growing states in India. High level of correlation between average crop yield and average fertilizer use per hectare confirm the direct effect of fertilizer use on crop production. However, fertilizer use alone cannot be credited for increase in crop yields in the progressive states as other factors like irrigation, plant protection, crop varieties and weather also influence crop growth and performance.

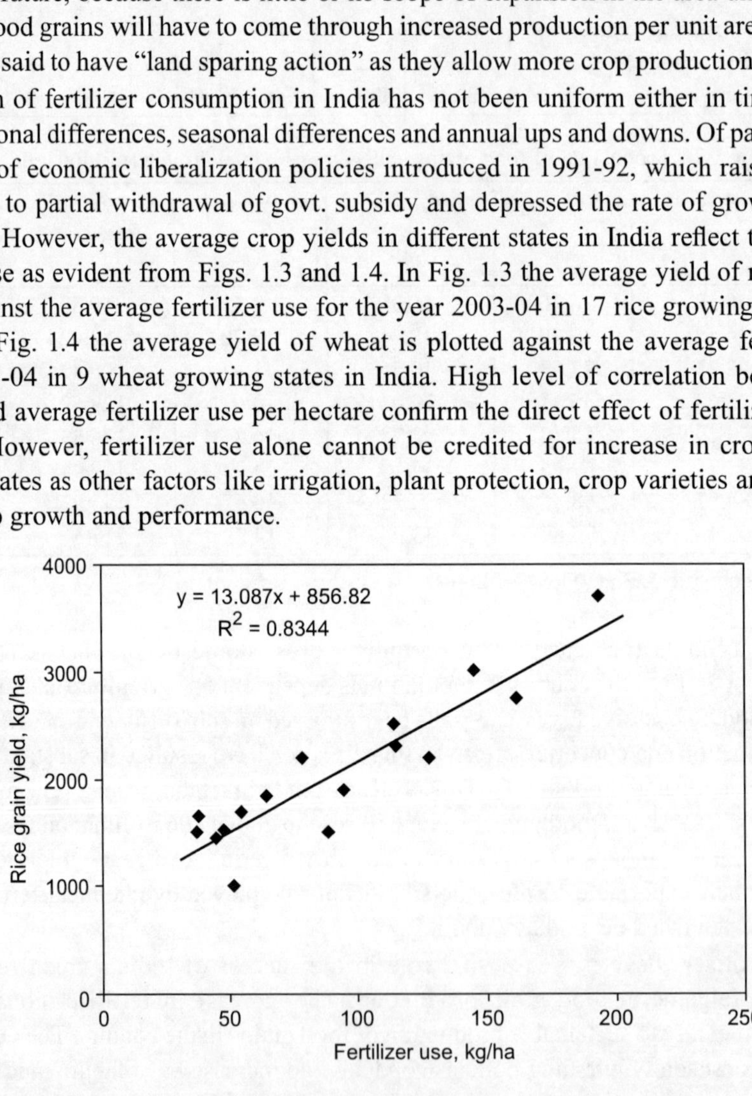

Fig. 1.3 Relationship between average fertilizer use and rice yield in 17 rice growing states in India (2003-04)

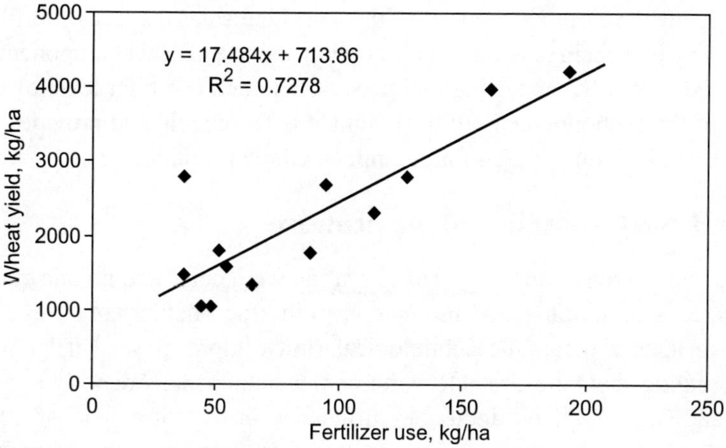

Fig. 1.4 Relationship between average fertilizer use and wheat yield in nine wheat growing states in India (2003-04)

FERTILIZERS AND AGRICULTURAL DEVELOPMENT

Sustainable Agricultural Development

The optimal use of fertilizers is an important element, together with optimal use of other available sources of plant nutrients, the efficient use of water, the avoidance of harmful emissions to air and water and other best practices that are essential for world agriculture to be sustainable.

Mineral fertilizers improve farming efficiency and water conservation. Mineral fertilization provides for nutrient balance and optimum soil productivity to stimulate healthy, vigorous crop growth. Fertilizers allow for quick canopy cover and establishment of efficient root systems and more aboveground plant growth. Fertilizers play a positive role in carbon dioxide (CO_2) assimilation by improving the fixation of CO_2 by crops, hence contributing to carbon sequestration.

Agricultural Development and Industry

Agriculture also spearheads industrial development by providing the necessary raw materials and by generating demand for industrial products. Hence, there is a close interdependence between agriculture and industry. Agriculture and agribusiness have the traditional strength of generating employment and income for rural uneducated population *vis-à-vis* information technology (IT), biotechnology and telecommunication which employ highly educated professionals in urban sector.

Development of agriculture and industry go hand in hand. A productive agriculture is necessary to supply the raw materials, the investment capital and the manpower for the industry while industry is necessary to supply many physical inputs such as fertilizers, crop protection chemicals, agricultural implements and transportation facilities. Fertilizer is an essential ingredient of the agricultural inputs required for enhancing agricultural production.

Experience in many developing countries of the world indicates that use of fertilizers by farmers is often the beginning of a chain reaction—the adoption and use of other components of the package of practices that are mutually supporting and necessary to achieve high level of production. This is especially true if the economic and political climate is favourable and provides the basis for the confidence that the innovations in use of more inputs will be profitable.

Fertilizers and Sustainability of Agriculture

In the mid-1960s, when projections of global starvation were common, no one questioned the role of mineral fertilizer in promoting food production in the food-deficit countries. On the contrary, fertilizer use was an integral part of the technological trinity: improved seed, irrigation, and fertilizer, responsible for bringing about the Green Revolution that helped many densely populated countries, including India and China, achieve food self-sufficiency in the short span of 20 to 25 years. In the early 1990s, however, fertilizer became a target of criticism mainly because of heavy use in the developed countries, where it was suspected of having an adverse impact on the environment through nitrate leaching, eutrophication, greenhouse gas emissions, and heavy metal contamination of soils and uptake by plants. Consequently, fertilizer use was mistakenly identified as an enemy of the environment. Although fertilizer use can contribute to environmental contamination unless managed properly, it is often an indispensable source of the nutrients required for plant growth and food production. Fertilizer use will remain an essential component of future strategies for ensuring food security and protecting the natural resource base. In fulfilling that role, however, fertilizer use should be approached differently in the future. Emphasis should be on growth with management rather than on growth per se, so that the broader goals of food security, agricultural growth, and environmental protection are not sacrificed.

The Limits of Organic Agriculture for Food Security

Organic agriculture, which eliminates the use of synthetic inputs (fertilizers, pesticides, herbicides) does not appear to be a feasible alternative. The consequences would be quite staggering if market demand for organic produce at a global scale increased substantially. The consequences are: a large amount of land that would have to be brought under rotation with legumes or under animal production to make up for the lack of mineral fertilizer. While organic agriculture does fill a niche market, its limits and its dangers, in terms of nutrient depletion should not escape attention.

The question is not whether but by how much fertilizer use will need to increase. At the World Food Summit in 1996, governments committed themselves to halving the number of hungry people by the year 2015. There is a direct link between that World Food Summit goal and fertilizer use. Possibly, it means an 8% increase in fertilizer applications compared to the "business-as-usual" scenario. That does not seem very much, but in terms of its tonnage, it is considerable. Enhanced fertilizer use to meet food production goals is particularly important in countries such as China and India, which make up a large proportion of the world population. But it may be even more important in Africa, where increase of 2.7% or more a year is needed in order to make up for nutrient losses, and in the humid tropics, where unfertilized annual cropping takes a heavy toll of soil organic matter.

Fertilizers are irreplaceable, particularly in the developing countries with poor soil fertility and high population pressure. Probably one of the most destructive systems in terms of environmental damage is unfertilized annual cropping in the humid tropics, with its high impact on soil organic matter and erosion. There is a need to look at systems in a much more integrated way. Such an approach to nutrient management is really important and should receive due attention of those concerned with fertilizers.

Recent results of the medium term study on wheat production comparing integrated farming system *vis-a-vis* organic and conventional farming at Leicester, England provide strong evidence for adopting integrated farming if high yields are to be achieved as organic farming produced much lower yields of wheat than conventional and integrated farming systems (Table 1.8). The **conventional farming** system gives the highest yields across the rotation, but the system is slightly less profitable than the integrated system. The **integrated farming** system gives wheat yields about 5% lower than the conventional system, but is more profitable overall due to savings on inputs. The **organic farming** system gives significantly lower wheat yields, about 60–70% that of the conventional and integrated systems. However, on a small scale its profitability may be equivalent to the conventional system due to additional subsidies and price premiums which can be as high as 100%.

Table 1.8 Comparison of wheat yields of three farming practices in Leicester, England (t ha^{-1})

Year	*Conventional farming*	*Integrated farming*	*Organic farming*
1994	5.67	6.59	5.09
1995	9.04	7.74	4.03
1996	10.3	9.78	6.66
1997	6.94	6.35	2.38
1998	6.53	6.24	3.38
Average	7.7	7.34	4.31

Fertilizers Spare the Land

Mineral fertilizers and land are substitutable in the sense that an increase in the use of fertilizers permits a reduction in the area of land cultivated, and *vice versa*. The use of mineral fertilizers has environmental costs but all farming, as most of man's activities, has an environmental impact. Overwhelmingly the evidence is that mineral fertilizers are necessary for the welfare of mankind. There are environmental risks but they are minor in relation to the benefits.

N.E. Borlaug (1997) stated: "*Take the cases of the United States, India and China as examples. In 1940, when relatively little inorganic fertilizer was used, the production of the 17 most important food, feed and fibre crops in the USA totalled 252 million tonnes from 129 million hectares. Compare these statistics with 1990, when American farmers harvested approximately 600 million tonnes from only 119 Mha – 10 Mha less than 50 years before. If the United States attempted to produce the 1990 harvest with the technology that prevailed in 1940, it would have required cultivating an additional 188 million hectares of land of similar quality. This theoretically could have been achieved either by ploughing up 73% of the nation's permanent pastures and rangelands, or by converting 61% of the forest and woodland area to cropland. In actuality, since many of these lands are of much lower productive potential than the land now in crops, it really would have been necessary to convert a much larger percentage of the pasture and rangelands or forests and woodlands to cropland. Had this been done, imagine the additional havoc from wind and water erosion, the obliteration of forests and extinction of wildlife species through destruction of their natural habitats, and the enormous reduction of outdoor recreation opportunities. Impressive savings in land use have also accrued to China and India through the application of modern technology to raise yields. Had the cereal yields of 1961 still prevailed in 1992, China would have needed to increase its cultivated cereal area by more than threefolds and India by about twofold, to equal their 1992 harvests. Obviously, such a surplus of agricultural land was not available.*"

Epilogue

Mineral fertilizers are an important resource for the humanity, as the world food production depends on use of mineral fertilizers to supplement the plant nutrients obtained from the soil. These are now indispensable for ensuring sufficient food production and preventing declines in soil productivity through nutrient depletion. The rapid increase in the world population and the consequent rise in food consumption have rendered fertilizers an integral part of the food supply chain. There is no alternative to mineral fertilizers in modern agriculture if we shall succeed in feeding the world.

However, current food production is putting serious strain on the environment, and agriculture is a source of pollution and environmental problems. Some of the issues such as nitrate leaching, eutrophication and consumption of non-renewable resources, originate with fertilizer use. These and related questions raise concern as to whether our agricultural practices are sustainable, and how we can cope with the needs of a growing population.

The challenge is to provide food for an increasing population while maintaining soil quality and taking care of a precious natural environment. One option to solving these pressing dilemmas is to improve fertilizing practices. This entails supplying plants with precise and balanced amounts of nutrients needed for their optimal growth and development, but avoiding excessive amounts that can be eventually lost into the environment.

Selected References

Greenwood, D.J. (1981). Fertilizer use and food production: world scene. *Fertilizer Research* **2**:33-51.

Prasad, R. (1999). Sustainable agriculture and fertilizer use. *Current Science*. **77**: 38-43.

Russel, D.A. and Williams, G.G. (1977). History of chemical fertilizer development. *Soil Sci. Soc. Amer. J.* **41**:260-265.

Stewart, W. M., Dibb, D. W., Johnston, A. E. and Smyth, T. J. (2005). The contribution of commercial fertilizer nutrients to food production. *Agron. J.* **97**:1-6.

Tiwari, K.N. (2002). Nutrient management for sustainable agriculture. *J. Indian Soc. Soil Sci.* **50**: 374-397.

CHAPTER

2

Fertilizer Feedstocks and Raw Materials

For production of various fertilizers, certain feedstocks and raw materials are required which are either available locally or they are transported from other places within the country or imported from global suppliers.

Raw Materials for Nitrogen Fertilizers

The most important raw material required for production of nitrogen fertilizers is ammonia which provides nitrogen to all the nitrogen fertilizers and nitrogen based compound fertilizers. Requirement of other materials depends on the choice of nitrogen fertilizer to be produced. For instance, urea production requires carbon dioxide and ammonia, whereas manufacture of ammonium sulphate requires sulphuric acid which is produced from sulphur.

Feedstocks for Ammonia

Ammonia is the key input required for manufacture of nitrogen fertilizers and nitric acid. It is produced by synthesis of nitrogen and hydrogen gases under high pressure and temperature. Nitrogen gas is available in abundance (78%) in the atmosphere. All the nitrogen required in ammonia synthesis is derived from air. The hydrogen gas is obtained from various raw materials such as natural gas, naphtha, oil or coal. These raw materials are called as "feedstock". The feedstocks for ammonia plants may broadly be grouped into three categories, viz. gaseous, liquid and solid.

Gaseous feedstocks: This category includes natural gas, refinery gas and coke oven gas. These feedstocks have lower calorific value but are the most suitable raw materials due to their high H_2/C ratio (2 for natural gas) and higher conversion efficiency.

Natural gas occurs either in association with crude petroleum or as non-associated gas. Accordingly, it is known as associated gas or free gas. The recoverable reserves of natural gas in India have been estimated as over 700 billion m^3. New locations of the gas reserves are also being discovered.

Natural gas is essentially methane (60-96%) with an impurity of ethane (4-40%). It is a preferred feedstock because it contains little or no obnoxious impurity. The cost is low and it releases H_2 by a simple reaction with steam in which approximately as much H_2 is produced from steam as from natural gas.

Refinery gas is a by-product of refinery and petroleum operations. It contains 80-90% H_2 and aliphatic hydrocarbons not heavier than butane (C_4H_{19}). Only very large refinery will however produce enough gas for use as feedstock for economically sized ammonia plant.

Coke oven gas is obtained during coking of coal, which is invariably done by steel plants. It contains 50-60% H_2, 22-28% CH_4, 5-9% CO, 4-12% N_2 and 2% CO_2. The gas must be purified before it is used as feedstock for ammonia production.

Liquid feedstocks: This category includes naphtha, heavy fuel oil, liquefied petroleum gas and water. These raw materials have higher calorific value (10340 kcal/kg of naphtha, 9829 kcal/kg of fuel oil) and high H_2/C ratio (0.85/1.1) resulting in higher thermal efficiency and low purification requirements.

Naphtha is a liquid petroleum distillate (specific gravity 0.71) with a final boiling point not higher than 215°C. Chemically, naphtha consists of hydrocarbons—paraffin 75%, naphthene 36% and aromatics 15%. It consists of mainly butane and higher hydrocarbons with little impurity of sulphur. Typical analysis of naphtha is given in Table 2.1.

Table 2.1 Typical analysis of naphtha

Property	*Value*
Specific gravity	0.70-0.72
Sulphur (ppm)	1500 (max.)
Lead (ppb)	5.0 (max.)
Chloride (ppm)	0.1 (max.)
R.O.E. (ppm)	5.0 (max.)
Hydrocarbons	
Paraffin (volume %)	75 (max.)
Olefin (%)	2
Naphthene (%)	36.0
Aromatics (%)	15.0
C/H ratio (by weight)	5.5-6.0
Boiling point (FBP)	180°C
Calorific value (kcal/kg)	10,500

Heavy fuel oil is a residue from refining operations. It is a viscous liquid (specific gravity 0.92) with relatively high sulphur content (2.0%). Normally, it does not compete economically with naphtha due to elaborate purification system required and higher investment cost. Typical analysis of fuel oil and low sulphur heavy stock (LSHS) from Indian refineries is shown in Table 2.2.

Table 2.2 Typical analysis of fuel oil and low sulphur heavy stock (LSHS) from Indian refineries

Property	*Fuel oil*	*LSHS*
Specific gravity	0.896-0.928	0.92
Sulphur (%)	1.57-2.07	0.91
Ash (%)	0.018-0.03	–
Carbon (%)	84.67-86.64	87-88
C/H ratio	6.33	7.72
Calorific value (kcal/kg)	9421-9932	10,360

Liquefied petroleum gas is mainly propane and butane separated by liquefaction from natural gas or refinery gas. Its impurities are olefins, COS and other sulphur compounds.

Use of water as a source of H_2 gas requires availability of abundant cheap electricity for its electrolysis. The fertilizer plant at Nangal, Punjab (National Fertilizers Ltd.) was initially planned to obtain H_2 from electrolysis of water but shortage of power in Bhakra grid compelled to switch over to liquid feedstock.

Solid feedstocks: This category includes coal, coke, lignite, etc. The solid raw materials have the lowest calorific value and the lowest H_2/C ratio. The thermal efficiency of conversion of these materials into synthesis gas ($3H_2 + N_2$) is poor and they require large and elaborate processing and purification systems to handle the oxides of carbon. The disposal of the ash is another problem.

India has a large coal reserve estimated at 211.6 billion tonnes (Table 2.3), of which 77% is located in three states of Orissa, Jharkhand and Chhattisgarh. The proven minable reserves are estimated at 72.7 billion tonnes. The coal mined from most of the Indian coalfields has relatively high ash content (38-44%). Typical analysis of non-coking coal is shown in Table 2.4. Also the process of gasification of coal or coke is not very flexible. For these reasons ammonia synthesis plants based on solid feedstocks are restricted to places where their cost is very low in comparison with liquid or gaseous raw materials.

In India, three large size coal based fertilizer plants have been established at Ramagundam (Andhra Pradesh), Talcher (Orissa) and Korba (Chhattisgarh).

Table 2.3 Coal reserves in India

State	*Reserve (million tonnes)*
Andhra Pradesh	13.5
Jharkhand	69.1
Madhya Pradesh and Chhattisgarh	43.4
Maharashtra	7.1
Orissa	50.4
Uttar Pradesh	1.1
West Bengal	25.9
Others	1.1
Total	211.6

Table 2.4 Analysis of non-coking coal

Constituent	*Value (%)*
Moisture	85–1.76
Ash	38.8–43.7
Sulphur	–
Volatile matter	16–19
Fixed carbon	35.8–43.3
Calorific value (kcal/kg)	3994-4395

Raw Material for Phosphate Fertilizers

The essential raw material for phosphate fertilizer is rock phosphate which provides phosphorus or P_2O_5. Sulphuric acid, nitric acid or electric energy are other essential materials or inputs required for manufacture of phosphate fertilizers from rock phosphate. Sulphuric acid is produced from sulphur whereas nitric acid is obtained from the oxidation of ammonia.

ROCK PHOSPHATE

Natural Phosphate Deposits

The original source of phosphorus in the manufacture of phosphate fertilizers in early days was bones, but the supply was soon exhausted. Today, rock phosphate is the only important raw material for phosphate fertilizers.

The earth is richly endowed with phosphate. There are two types of phosphate deposits: igneous and sedimentary. Both have essentially the same phosphate mineral—calcium phosphate of apatite group. Igneous deposits occur in veins and replacement bodies commonly associated with magnetite. Such deposits are not much extensive in size and number and not so easily mined and processed as sedimentary deposits. Major igneous deposits are located in Russia, South Africa, Brazil, Uganda, Botswana and Sri Lanka. About 15% of the world phosphate production is from igneous deposits.

Important phosphate deposits of the world are of sedimentary origin. These deposits are also referred to as *phosphorite*. About 95% of the world phosphate reserve and 82% of the commercial deposits are sedimentary. Major sedimentary deposits are located in USA, Australia, Togo, Morocco, Algeria, Jordan and Asian countries.

Guano is also sedimentary deposit which originated chiefly through accumulation of excrements of sea fowl and such deposits are largely confined to islands in southern latitudes. Some guano deposits are found in Chile, Peru, Seychelles and Philippines but much of the material has been exploited.

World Reserve of Rock Phosphate

In spite of the universal distribution of phosphate in nature, 99% of the world production of phosphate is contributed by 17 major countries and only three countries: China, USA and Morocco produce more than 60%. Estimates of worldwide reserves are placed at 50 billion tonnes (Table 2.5) and annual mine production of rock phosphate in 2006 was 145 million tonnes.

With increasing demand for phosphate fertilizers, the production of rock phosphate will have to be stepped up. Since phosphate is non-renewable natural resource, the exploitation of the deposits should be done in keeping with the future requirements.

Table 2.5 World reserve and production of rock phosphate (million metric tonnes)

Country	*Reserves*	*Reserve base*	*Production (2006)*	*% of world mine production*
China	6600	13,000	32	22.1
USA	1200	3400	30.7	21.2
Morocco	5700	21000	25.3	17.4
Russia	200	1000	11	7.6
Tunisia	800	600	8.4	5.8
Jordan	900	1700	6.4	4.4
Brazil	260	370	5.5	3.8
Syria	100	800	3.6	2.5
Israel	180	800	3.0	2.1
Egypt	100	760	2.74	1.9
South Africa	1500	2500	2.6	1.8
Australia	77	1200	2.05	1.4
Senegal	50	160	1.5	1.0
Togo	30	60	1.2	0.8
Canada	25	200	1.0	0.7
Other countries	800	2200	6.7	4.6
World total	18000	50,000	145	100

Phosphate Deposits in India

Considering the increasing need of phosphate fertilizers in India, the reserves of rock phosphates known to date are very meagre. The total reserve of phosphates of all categories in the country is estimated at 145 million tonnes. The major deposit of 79 million tonnes is found at Jhamarkotra near Udaipur in Rajasthan. Other deposits are located at Jhabua and Hirapur in Madhya Pradesh, Mussoorie in Uttarakhand and Purulia in West Bengal. Of these, only the deposits at Jhamarkotra (Udaipur), Jhabua and Hirapur are of high quality (30% P_2O_5), whereas all other deposits have poor quality ore with low phosphorus content (15-20% P_2O_5). Small quantities of rock phosphate deposits have also been discovered at Kashipatnam (Andhra Pradesh), Singhbhum (Jharkhand), Tiruchirapalli (Tamil Nadu) and Lalitpur (Uttar Pradesh).

At present, mining is being done at Udaipur, Jhabua and Purulia with total production of about 1.5 million tonnes annually.

Mining of Rock Phosphate

Depending on the thickness of overburden and depth of the phosphate matrix either strip mining or underground mining is followed.

Strip mining: This is a surface mining method which is also called as open cut method. In this method overburden is first removed by mobile dragline excavators or power shovels or caterpillar powered scrapers. The phosphate matrix is dug by separate dragline scraper and deposited in sumps where it is broken up.

Underground mining: When phosphate matrix lies beneath a very deep or hard rock overburden tunnels are made into the hillside. The phosphate matrix is either drilled or blasted and taken out by conveyors. Underground mining is quite expensive, hence less popular.

Beneficiation of Rock Phosphate

Fertilizer industry requires a high grade rock phosphate in order to produce fertilizers of desired quality. For instance, superphosphate industry would not use a rock phosphate containing less than 30% P_2O_5 (13.1% P). Generally, rock phosphate with +31% P_2O_5, Fe_2O_3 and Al_2O_3 below 2%, silica below 5%, fluorine 3.5%, MgO 0.5% and chlorine below 100 parts per million (ppm) is considered as good raw material for production of phosphate fertilizers. If the rock phosphate obtained from the mine is of high grade, it is shipped as such to the fertilizer industry. It is however often poor in quality containing 15-20% P_2O_5 and a number of impurities. Such ores are upgraded by beneficiation process. Beneficiation is a process of ore dressing in order to remove impurities and thereby to improve the grade of the ore. The impurities are collectively referred to as *gangue*. Main impurities are iron and aluminium oxides, calcium carbonate, silica, clay, etc. These impurities not only lower the grade of the ore but also consume substantial quantity of acid or energy while converting the rock phosphate into finished products. For example, iron and aluminium oxides would cause formation of undesirable phosphate compounds in the product. If silica content is too high, the requirement of power for grinding would be more. Presence of free carbonates (usually $CaCO_3$) would increase the acid consumption. For these reasons, the impurities must be reduced to the acceptable limits by the beneficiation process. Beneficiation would also ensure consistency of the product grade. There are three main processes of beneficiation of rock phosphate: washing, flotation and calcination.

Washing: Crushing or grinding is the first step which reduces the size of pebbles and nodules, and releases impurities. This is followed by water washing to remove soluble impurities such as sodium chloride, free lime and fine materials like clays, silica and other debris. A sizing operation may follow the washing step. Wet sizing as well as dry sizing is practised.

Flotation: Generally, two stages of flotation are used to obtain high grade rock phosphate. In the first stage, deslimed feed (after removal of clay) is subjected to fatty acid or amine flotation for phosphate enrichment. This is followed by amine or cationic flotation of the residual silica. The flotation agent selectively films the desired particles leaving others in the suspension. In the flotation cell, air is drawn by agitation and air bubbles attach to the filmed particles which then float to the surface. The froth is mechanically skimmed off.

Flotation is not a clean cut process. Flotation depressants are sometimes added to increase the selectivity of the flotation agent for the desired component. Starch has been used to depress the flotation of calcium carbonate and sodium metasilicate has been used to depress flotation of silica and magnetite. Phosphate recovery by flotation process is usually 60-80%. Thus, considerable quantity of phosphate is lost in tailing and slimes.

Calcination: In this process rock phosphate is heated with steam to the temperature (400-1700°C) which is below the melting point. Heating liberates fluorine which escapes into air. Carbonates decompose and organic contaminants are oxidized and eliminated.

The calcined ore can be recycled through washing, flotation or any other method if it fails to meet the desired ore specification.

Characteristics of Rock Phosphate

Rock phosphate is essentially a trade name covering a wide variety of phosphate ores having diverse mineral composition, colour and texture. They all, however, have a common characteristic, i.e. they contain one or more of the apatite group minerals. In igneous deposits, the principal mineral is fluorapatite, $Ca_{10}(PO_4)_6F_2$ whereas in sedimentary marine deposits, it is usually a substituted carbonate fluorapatite or francolite having formula $Ca_{10-(a+b)}Mg_aNa_b(PO_4)_{6-x}(CO_3)_xF_{0.4x}(OH, F)_2$. In commercial rock phosphate samples apatite content is usually from 72 to 92 per cent. Non-phosphate impurities are calcite, dolomite, quartz, clays, and hydrated iron-aluminium oxides. Composition of some typical samples of rock phosphate is shown in Table 2.6.

Rock phosphate is marketed in ground fine form having gray, brown or brown-black colour with rough texture.

Table 2.6 Typical analysis of imported and Indian rock phosphates (on dry basis)

Rock phosphate source	P_2O_5	CaO	SiO_2	F	CO_2	Al_2O_3	Fe_2O_3	MgO
Imported rocks								
Florida rock	32.5	48.8	4.7	3.9	3.8	1.0	1.1	0.3
Morocco rock	33.0	51.6	3.3	2.8	5.3	0.1	0.2	–
Trans-Jordan rock	32.3	51.6	4.7	3.2	5.1	0.6	0.3	–
Syria rock	31.9	51.8	4.0	3.7	5.1	0.8	0.2	–
Indonesian rock	37.4	52.8	0.7	0.3	4.2	0.6	0.2	–
Algerian rock	31.4	52.0	2.8	3.8	3.7	0.6	0.6	–
Brazilian rock	34.5	50.4	3.5	3.7	4.2	1.2	0.6	1.7
Thailand rock	28.2	49.9	2.1	0.9	10.1	1.3	0.7	–
Indian rocks								
Jhamarkotra (H.G.)	32.0	46.0	9–20	3.0	–	1.5	0.5	1.5
Jhamarkotra (L.G.)	18–20	38–42	6–8	1.2	–	1.2	-	7-10
Jhabua	24.0	42–48	14–18	2.5–3.0	1.0–1.6	1.8-3.5	1.5-2.5	0.5-0.75
Hirapur	27–29	42–52	5–8	3.0–3.5	0.3–1.0	0.5-3.0	1.5-8.0	0.1-1.0
Purulia	19.5	24.9	15.7	–	–	5.0	20.8	1.9

SULPHUR

Sulphur is the essential raw material for sulphuric acid. About 80 to 85% of the world's sulphur production is used to manufacture sulphuric acid. It is also used for direct application as soil amendment to alkali (sodic) lands and source of nutrient to sulphur deficient soils.

Sources of Sulphur

There are two principal sources of elemental sulphur:

- the natural deposits of sulphur; and
- sulphur recovered from sour gas, petroleum refining or metallurgical operations.

Sulphur Deposits in the World

Natural deposits of elemental sulphur are found in two forms. Either they occur in dome-like formations in association with gypsum deep below the earth surface or they are deposited on the earth surface by volcanic activity. Dome sulphur deposits occur along the coast of the Gulf of Mexico in USA and Mexico. Elemental sulphur is found near hot springs and volcanic regions in many parts of the world. Such volcanic deposits are currently mined in Indonesia, Chile and Japan. Sicily is also famous for its sulphur mines. Significant deposits of elemental sulphur also exist in evaporates in Eastern Europe and western Asia.

The largest world reserves of sulphur are in Iraq, Canada, Spain, China, Poland and the United States. It is estimated that the world resources in terms of sulphur associated to oil, natural gas, metallic sulphides, evaporates and deposits of volcanic origin may amount to 5 billion tonnes.

Method of Extraction

The world production of sulphur in 2006 was approximately 66.5 million tonnes. The major producers are the United States (14%), followed by Canada (13.5%), Russia (10.5%) and Japan (5%).

There are two principal methods of producing elemental sulphur:

- mining of the natural deposits; and
- recovery from sour gas, petroleum refining or metallurgical operations.

Sulphur Mining

Sulphur is extracted from the natural deposits mainly by two processes: the Sicilian process and the Frasch process. The Sicilian process, which was first used in Sicily, was used in ancient times to get sulphur from rocks present in volcanic regions. In this process, the sulphur deposits are piled and stacked in brick kilns built on sloping hillsides, and with airspaces between them. Then powdered sulphur is put on top of the sulphur deposit and ignited. As the sulphur burns, the heat melts the sulphur deposits, causing the molten sulphur to flow down the sloping hillside. The molten sulphur then can be collected in wooden buckets.

The second process used to obtain sulphur from underground deposits is the Frasch process. In this process, three concentric pipes are used: the outermost pipe contains superheated water, which melts the sulphur, and the innermost pipe is filled with hot compressed air, which serves to create foam and pressure. The resulting sulphur foam is then expelled through the middle pipe. The Frasch process produces sulphur with 99.5% purity, which needs no further purification. The sulphur produced by the Sicilian process must be purified by distillation.

Recovery from Sour Gas

Elemental sulphur is also recovered from sour gas. Sour gas is a term applied to natural gas containing appreciable quantities of hydrogen sulphide (H_2S). Before sour natural gas can be sold, it must be sweetened to remove rotten-egg odour of hydrogen sulphide. Depending on the amount of hydrogen sulphide in the sour gas, in some wells sulphur is a by-product, in some it is co-product and in others it is the main product. Natural gas deposits have been found that are very rich in hydrogen sulphide, up to 87%. The hydrogen sulphide is removed from the sour gas by an absorbent solution which is sent to reactivator to remove hydrogen sulphide (acid gas). Sulphur is recovered from the acid gas by oxidation: either total combustion or one-third to two-thirds method. Total combustion process is simple oxidation of hydrogen sulphide in a combustion chamber giving elemental sulphur by the reaction:

$$H_2S + 1/2O_2 = 1/6S_6 + H_2O$$

In one-third to two-thirds method, hydrogen sulphide gas is divided into one-third and two-thirds streams and the one-third stream is burned in the furnace of steam generator to produce sulphur dioxide. Sulphur dioxide is then recombined with two-thirds stream of hydrogen sulphide and the gas mixture goes to a bauxite catalyst reactor where the reaction between hydrogen sulphide and sulphur dioxide gives liquid sulphur. A two-stage coalescer is used to extract more liquid sulphur.

$$H_2S + 3/2O_2 = SO_2 + H_2O$$
$$2H_2S + SO_2 = 1/6S_6 + 2H_2O$$

Sulphur is also a by-product of refining operations. In crude oil, sulphur is present in organic combinations which upon cracking or hydrogenation gives hydrogen sulphide. Sulphur is recovered from hydrogen sulphide by the same process as from sour gas.

Recently, sulphur contamination in crude oil has provided a major source of sulphur. In 1990 mined sulphur made up for around one-third of the world's total sulphur supply, it now makes up less than one per cent, with sulphur extracted from crude oil and natural gas making up for around 98% of the world's total sulphur production.

Characteristics of Elemental Sulphur

Elemental sulphur is light yellow, inert crystalline solid. It is odourless and insoluble in water. Under certain conditions, sulphur dust and air form a potentially explosive mixture.

Sulphur oxidizes slowly on contact with soil, air or the atmosphere. Sulphur ignites at 250°C. When burnt, it produces sulphur dioxide, a pungent gas, which is widely used in the manufacture of sulphuric acid. About 75-85% of elemental sulphur is used for this purpose only.

POTASH

Raw Materials for Potash Fertilizers

The natural salt deposits of potash and brines in the salt lakes provide the raw material for processing into potash fertilizers. About 95% of the global potash supply is used for fertilizer industry.

Potash Resources and Reserves

Potassium occurs abundantly in nature. It is the 7th most common element in the earth's crust which contains 2.4% potassium. Certain clay minerals associated with heavy soils are rich sources of potassium, containing as much as 17% potassium. Sea water typically contains 390 mg K/L representing a huge total amount of the element globally.

Large potash bearing rock deposits occur in many regions of the world deriving from the minerals in ancient seas which dried up millions of years ago. Potassium deposits occur as beds of solid salts at varying depths below earth's surface and also as brines in dying lakes and sea. At present underground potassium salt deposits are the major raw materials for fertilizer production and account for over 90% of the world potassium reserve. Potash for fertilizer is mainly derived from this potash rock, requiring only separation from the salt and other minerals and physical grading into a form suitable for fertilizer manufacture or farm spreading.

Principal potash deposits and annual production in the world are listed in Table 2.7. Total potash deposits reserves are estimated as 8265 billion tonnes with a reserve base of 16545 billion tonnes. The largest reserve of potash salts is found in Canada. Nearly half of the world potash reserve occurs in Canada alone. Russia has also a large deposit of potash and ranks second in the world. So far, there are no known deposits of potassium in India, though it is fourth largest consumer of potassium fertilizers in the world.

Table 2.7 World mine production, reserves and reserve base potash

Country	*Mine production in 2006 (million tonnes)*	*Reserve (million tonnes)*	*Reserve base (million tonnes)*
United States	1,200	90,000	300,000
Belarus	4,000	750,000	1,000,000
Brazil	405	300,000	600,000
Canada	10,200	4.400,000	9,700,000
Chile	350	10,000	50,000
China	700	8,000	450,000
Germany	3,660	710,000	850,000
Israel	2,100	40,000	580,000
Jordan	1,200	40,000	580,000
Russia	5,300	1,800,000	2,200,000
Spain	500	20,000	35,000
Ukraine	65	25,000	30,000
United Kingdom	600	22,000	30,000
Other countries	–	50,000	140,000
World total	30,280	8,265,000	16,545,000

Thickness of potassium containing bed varies from a few centimetres to several metres but generally the beds of one metre or more thickness are mined. Commercially important minerals of potassium are listed in Table 2.8. Sylvinite is by far the most important ore of potassium. Carnallite, langbeinite and kainite are lower in K content and relatively difficult to process. Other minerals such as shoenite, glaserite and syngite are of little significance commercially.

Table 2.8 Commercially important potash minerals

Name of mineral	*Formula*	K_2O (%)	*Remarks*
Sylvite	KCl	63.2	Principal ore mineral
Sylvinite	KCl + NaCl	10-35	Principal ore
Carnallite	$KCl\ MgCl_2.\ 6H_2O$	16.9	Ore mineral
Kainite	$4KCl\ 4MgSO_4.11H_2O$	19.3	Ore mineral
Langbeinite	$K_2SO_4.\ 2MgSO_4$	22.7	Ore mineral
Polyhalite	$K_2SO_4\ MgSO_4.2CaSO_4.\ 2H_2O$	15.5	Ore contaminant
Niter	KNO_3	46.5	Ore mineral

Mining Potassium Ores

The method of mining potassium ores depends on the nature of deposit. The process of mining is complicated by the great depth at which the salts occur. Two principal methods of mining, viz. shaft mining and solution mining are employed for solid ores.

Shaft mining: This method is used for the solid ore which lies in beds not deeper than 1100 metres. Usually room and pillar mining pattern is used for raw ore extraction. In this method, rectangular rooms are mined out and pillars of the ore are left to support the overburden. As such only 50-60% of the ore is recovered in the primary mining. Subsequent second stage mining or 'pillar robbing' results in recovery of about 90% of the ore. Conventional coal mining techniques involving face undercutting, drilling and blasting were commonly used till recently when continuous mining machines replaced the conventional methods. These machines cut the ore directly from mine face.

Working conditions in a potash mine are much superior to those encountered in other types of mines. Potash mines are dry, well ventilated and clean with comfortable temperature of about 24°C. The reddish pink colour of crystals provides a pleasant atmosphere.

Solution mining: Where depth is too high, it is not safe and economic to use shaft mining system. Potassium salts are recovered from such deposits by solution mining method, which consists of the following operations:

(i) Drilling wells upto the depth of potassium deposit
(ii) Injection of a brine of controlled NaCl-KCl composition and temperature to dissolve the potassium salt in the deposit
(iii) Pumping of the resultant solution from the underground cavity to surface
(iv) Concentration and removal of desired salt, viz. potassium chloride
(v) Adjustment of the composition of remaining brine for recirculation to the first step.

Solution mining permits a greater recovery of potassium deposit than shaft mining but requires drilling of several wells and more expensive surface refinery to crystallize potassium chloride from a large volume of brine.

Brine Sources of Potash

Relatively high concentration of potassium salts is present in brines of some salt lakes, Dead Sea and underground wells. Commercially important salt lakes for potassium extraction are Searle's Lake, California; the Great Salt Lake, Idaho; and the Dead Sea. Brines contain mainly potassium chloride with several impurities such as sodium chloride, sodium sulphate, calcium chloride, magnesium chloride, etc. Great Salt Lake of Utah and Searle's Lake of California are two important salt lakes providing brines for extraction of potassium chloride. Typical brine composition of Searle's Lake, California (USA) is given in Table 2.9.

Table 2.9 Typical brine composition of Searle's Lake, California

Component	*Formula*	*Upper structure*	*Lower structure*
Potassium chloride	KCl	5.02	2.94
Sodium chloride	NaCl	16.06	15.51
Sodium sulphate	Na_2SO_4	6.75	6.56
Sodium carbonate	Na_2CO_3	4.8	6.78
Sodium tetraborate	$Na_2B_4O_7$	1.63	1.96
Potassium bromide	KBr	0.12	0.08
Sodium sulphide	Na_2S	0.08	0.04
Phosphates (as P_2O_5)		0.07	0.04
Lithium (Li_2O)		0.15	0.006

Dead Sea is also a salt lake which lies in the Jordan Rift Valley between Israel and the West Bank in the west, and Jordan in the east. It is 422 metres below sea level and is 378 m deep, the deepest hypersaline lake in the world. It is also one of the world's saltiest bodies of water, with 33.7 per cent salinity, which is 8.6 times saltier than the ocean. The Dead Sea is 67 kilometres long and 18 kilometres wide. Typical brine composition of Dead Sea is given in Table 2.10.

Table 2.10 Typical brine composition of Dead Sea

Component	*Concentration (g/kg)*
Chloride	181.4
Bromide	4.2
Sulphate	0.4
Calcium	14.1
Sodium	32.5
Potassium	6.2
Magnesium	35.2

Selected References

FAI (2001). *Handbook on Fertilizer Technology*. 6th ed., Fertiliser Association of India, New Delhi. pp. 273.

UNIDO/IFDC, Ed. (1998). *Fertilizer Manual*, 3rd ed., Kluwer Academic Publishers, Dordrecht, the Netherlands.

CHAPTER

3

Acids Used for Fertilizer Production

Three acids, namely, nitric acid, sulphuric acid and phosphoric acid are used extensively in the fertilizer industry for production of various products. The production of these acids is discussed briefly.

NITRIC ACID

The industrial uses of nitric acid are extensive. It is used in fertilizers, synthetic fibres, explosives, dyes, synthetic rubber and pharmaceuticals. Today, the primary outlet of nitric acid is fertilizers where about 75% of the acid is consumed.

Raw Materials

Ammonia and air are required for the manufacture of nitric acid. Depending on the specific process ammonia consumption is 0.28-0.29 tonne per tonne nitric acid produced.

Method of Manufacture

Before the establishment of synthetic ammonia industry, nitric acid was produced by the reaction of sodium nitrate with sulphuric acid. The availability of inexpensive nitrogen in ammonia provided a more economical process for the production of nitric acid. Today, nitric acid is entirely produced by the ammonia oxidation process. The superheated ammonia gas is mixed with air and sent to a converter containing platinum-rhodium catalyst gauzes where ammonia oxidation takes place at a temperature of 850 to 950°C. In the converter, ammonia is first converted to nitric oxide (NO) which is then converted to nitrogen dioxide (NO_2) with the help of secondary air. The process water absorbs nitrogen dioxide to form nitric acid.

$$\underset{\text{Ammonia}}{4NH_3} + \underset{\text{Oxygen}}{5O_2} = \underset{\text{Nitric oxide}}{4NO} + \underset{\text{Water}}{H_2O}$$

$$\underset{\text{Nitric oxide}}{2NO} + \underset{\text{Oxygen}}{O_2} = \underset{\text{Nitrogen dioxide}}{2NO_2}$$

$$\underset{\text{Nitrogen dioxide}}{3NO_2} + \underset{\text{Water}}{H_2O} = \underset{\text{Nitric oxide}}{NO} + \underset{\text{Nitric acid}}{HNO_3}$$

The reactions are exothermic in nature and once started are self-sustaining.

Recently, there have been revolutionary developments in the nitric acid technology. In contrast to the operations at atmospheric pressure earlier, now the oxidation and absorption processes operate at elevated pressure of 5-10 bars.

The acid is 50 to 65% strength and is stored in stainless steel vessels. It can be used in fertilizer industry.

For production of constant boiling point nitric acid (68% strength) further concentration by extractive distillation with concentrated sulphuric acid or magnesium nitrate is carried out.

Product Characteristics

Nitric acid is water coloured to light yellow liquid with a specific gravity of 1.42 at 15.5°C. It is completely soluble in water. It is a strong acid and a strong oxidizing agent. It is extremely reactive with metals (except stainless steel). It tends to darken in sunlight and reacts vigorously with plant and animal tissue.

SULPHURIC ACID

Sulphuric acid is an important heavy chemical that has a variety of industrial uses. Fertilizer industry is the major user of sulphuric acid for acidulation of rock phosphate and production of ammonium sulphate. Other users are steel works, rayon and staple fibre, alum and other inorganic chemicals, petroleum refining, etc. About 80 to 85% of the world's sulphur production is used to manufacture sulphuric acid. Half of the world's sulphuric acid production is used in fertilizer production, mainly to convert insoluble phosphates to water-soluble forms.

Raw Materials

Elemental sulphur, pyrites and sulphides are the basic raw materials for production of sulphuric acid. It is also produced as a by-product of smelting operations. In India, most of the factories use elemental sulphur for production of sulphuric acid whereas some plants use smelter gases from copper or zinc smelters. Only one factory is based on pyrites. About 1 tonne of sulphur is needed to produce 3 tonnes of sulphuric acid—enough to manufacture about 2 tonnes of merchant grade phosphoric acid (50—54% P_2O_5). More than 2 tonnes of diammonium phosphate (DAP) can be manufactured from 2 tonnes of P_2O_5 and the original tonne of sulphur.

Method of Manufacture

There are two processes of production of sulphuric acid: chamber process and contact process. The chamber process is the oldest process and is now obsolete.

In the contact process elemental sulphur (99.5% purity, free from arsenic, chloride, and fluoride) is burnt in an excess of dry air at a temperature of 950-1100°C to form sulphur dioxide (SO_2). The gases containing SO_2 are cooled to 450-500°C and then passed over vanadium pentaoxide catalyst charged in a converter. The oxidation of SO_2 to sulphur trioxide (SO_3) is carried out in a series of three or more beds of the catalyst. The gases are cooled further and sent to absorption tower where SO_3 is absorbed in recirculating sulphuric acid. The recirculating acid is diluted with water and temperature is maintained at 70-90°C by cooling. The final product is 97-98% strength. Unabsorbed gases are vented to the atmosphere.

$$S + O_2 = SO_2$$

$$\underset{\text{Pyrites}}{4FeS_2} + 11O_2 = 8SO_2 + 2Fe_2O_3$$

$$\underset{\text{Sphelerite}}{2ZnS} + 3O_2 = 2SO_2 + 2ZnO$$

Sulphuric acid containing excess SO_3 is known as oleum or fuming sulphuric acid. A flow diagram for contact process of sulphuric acid is shown in Fig. 3.1.

Product Characteristics

Sulphuric acid is heavy, colourless liquid with specific gravity of 1.834. It is very strong acid. Common impurities are arsenic and lead. It is stored and transported in steel tanks or lead-lined tanks.

PHOSPHORIC ACID

Fertilizer industry is the major user of orthophosphoric acid, commonly called phosphoric acid. Some quantity of phosphoric acid is used in the detergent, foodstuff and other industries. Two types of phosphoric acid named after the process of production, namely, wet process and furnace process are produced. They differ in concentration and purity.

Raw Materials

The requirement of raw materials depends on the process selected for production of phosphoric acid.

1. For wet process, a high grade rock phosphate (greater than 30% P_2O_5, free from lime and low quartz, iron and aluminium) and sulphuric acid are the raw materials. For production of each tonne of phosphoric acid 3.25 tonnes rock phosphate and 2.75 tonnes sulphuric acid are required.
2. For furnace process, rock phosphate, silica pebble, coke and electric power are needed.

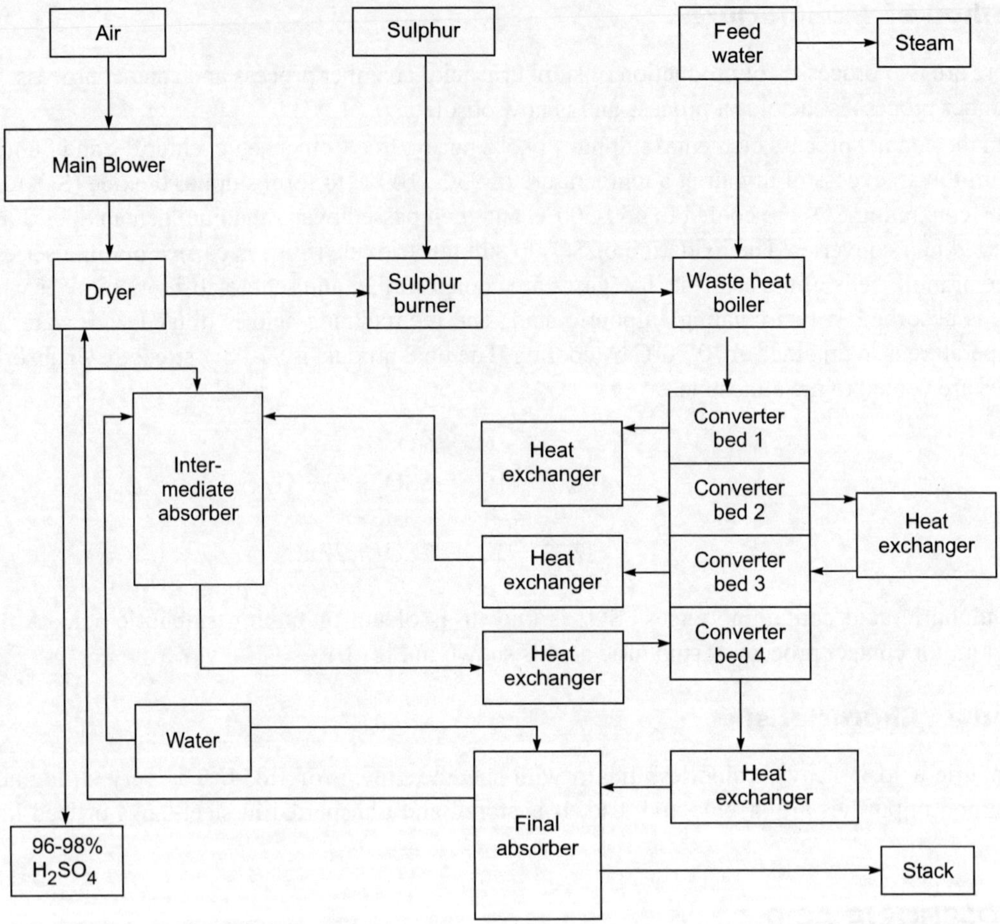

Fig. 3.1 Flow diagram for double contact double absorption process of sulphuric acid

Method of Manufacture

There are two methods of commercial production of orthophosphoric acid: wet process and furnace process. Wet process is commonly used for production of phosphoric acid in the fertilizer industry.

Wet Process

The four steps involved in the production of phosphoric acid are as follows:

(i) Acidulation of finely ground rock phosphate with sulphuric acid to form phosphoric acid and gypsum.

$$\underset{\text{Rock phosphate}}{Ca_{10}(PO_4)_6F_2} + \underset{\text{Sulphuric acid}}{10H_2SO_4} + 10H_2O = \underset{\text{Phosphoric acid}}{6H_3PO_4} + \underset{\text{Gypsum}}{10CaSO_4.2H_2O} + \underset{\text{Hydrogen fluoride}}{2HF}$$

(ii) Separation of gypsum from the acid by filtration

(iii) Countercurrent washing of gypsum to remove adhering phosphoric acid
(iv) Concentration of the acid by vacuum evaporation

A large quantity of by-product gypsum is produced at the rate of 5 tonnes per tonne phosphoric acid. It is disposed off in dry or slurry form. There are two processes for manufacture of wet process phosphoric acid, namely, dihydrate process and hemihydrate process. A combination of these two processes is also used.

Dihydrate process: This is the most popular process which is more flexible and reliable than the other process. This involves reaction between ground rock phosphate and sulphuric acid below 80°C. The product acid of 26-32% P_2O_5 concentration is filtered from the gypsum ($CaSO_4.2H_2O$) crystals. The filter cake is washed with countercurrent water flow on the filter and the dilute acid obtained from the filter is returned to the reactor. The phosphoric acid is evaporated to 40% in single stage and further evaporated to produce 50-54% P_2O_5 grade acid. Typical flow diagram for wet process phosphoric acid is shown in Fig. 3.2.

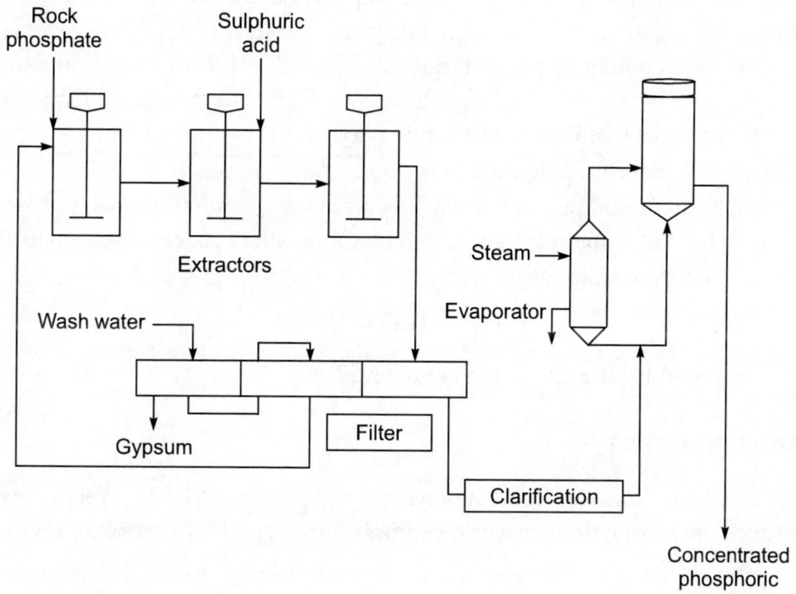

Fig. 3.2 Typical flow diagram for wet process phosphoric acid

Hemihydrate process: This process produces phosphoric acid of better purity at a lower cost. This involves reaction between rock phosphate and sulphuric acid above 80°C. The operating conditions are so selective that calcium sulphate hemihydrate ($CaSO_4.H_2O$) is crystallized during the reaction instead of dihydrate form (gypsum). The slurry is filtered to get phosphoric acid of 40 to 52% concentration saving considerable energy required for concentration by evaporation. One disadvantage of this process is the difficulty in filtration due to small size of crystals.

Recent developments in wet process phosphoric acid production have led to three-stage recrystallization process in which it is possible to change the form of calcium sulphate twice during the process and effect a P_2O_5 recovery of 99%.

Furnace Process

In this process elemental phosphorus is produced by mixing sand, coke and rock phosphate in an electric furnace at about 1600°C. Coke serves as source of carbon required for the reduction of fluorapatite, the phosphate mineral present in rock phosphate and sand is a source of silica which reacts with lime (CaO) formed during the reduction stage. At the temperature existing in the furnace, phosphorus as well as some of fluorine and carbon monoxide are liberated as gases. The main reaction occurring in the furnace is

$$\underset{\text{Fluorapatite}}{Ca_{10}(PO_4)_6F_2} + \underset{\text{Silica}}{3SiO_2} + \underset{\text{Carbon}}{5C} = \underset{\text{Calcium silicate}}{3(CaO.SiO_2)} + \underset{\text{Phosphorus}}{P_2} + \underset{\text{Carbon monoxide}}{5CO} + \underset{\text{Fluorine}}{F_2}$$

The furnace gases containing the phosphorus vapour are passed through a dust catcher—an electronic Cottrell precipitator unit while maintaining a temperature of 290°C to prevent condensation of phosphorus. The gases containing phosphorus, fluorine and carbon monoxide are then passed into a condenser where water is sprayed counter to the flow of gases to reduce the temperature to allow condensation of phosphorus to a liquid. The liquid phosphorus is drained into a storage tank where it is kept under water to prevent its burning.

The liquid phosphorus is pumped to burner; it is mixed with air and burned to form phosphorus pentaoxide (P_2O_5). The P_2O_5 vapour is then sent to hydrator where water is sprayed to allow reaction between P_2O_5 and water to form phosphoric acid.

$$P_2O_5 + 3H_2O = 2H_3PO_4$$

The acid concentration is 80-85% which contains 52-60% P_2O_5.

Product Characteristics

The phosphoric acid is a colourless liquid having specific gravity of 1.574. Wet process phosphoric acid has sometimes greenish colour because of impurities, primarily vanadium and is also known as green acid.

It contains 54% P_2O_5 (23.6% P) as orthophosphate (H_3PO_4). Furnace process phosphoric acid also known as white acid has high purity in contrast to the wet process acid which contains several impurities but is less expensive. For economic considerations, wet process phosphoric acid is used for production of solid fertilizers such as ammonium phosphate, triple superphosphate. The furnace phosphoric acid is preferred for production of liquid fertilizers.

SUPERPHOSPHORIC ACID

Superphosphoric acid is actually anhydrous phosphoric acid containing 72-76% P_2O_5 in comparison to 54% P_2O_5 in orthophosphoric acid. It is actually a mixture of orthophosphoric acid and polyphosphoric acid. The need for this acid arose from search for a high analysis phosphorus source for liquid mixed fertilizers in which use of wet process phosphoric acid was found unsuitable as it caused the precipitation of some compounds due to presence of impurities like iron and aluminium oxides in it. Alternately, use of white phosphoric acid (furnace process) was too expensive.

Method of Manufacture

Superphosphoric acid is produced by controlling the water used for hydration of phosphorus vapour in the hydrator in the furnace process. It can also be produced from wet process phosphoric acid by further concentration by evaporation.

Production of superphosphoric acid involves polymerization or condensation of orthophosphoric acid molecule by dehydration (loss of water) as shown below:

$$\underset{\text{Orthophosphoric acid}}{2H_3PO_4} = \underset{\text{Pyrophosphoric acid}}{H_4P_2O_7} + H_2O$$

$$\underset{\text{Orthophosphoric acid}}{2H_3PO_4} + \underset{\text{Pyrophosphoric acid}}{H_4P_2O_7} = \underset{\text{Polyphosphoric acid}}{H_5P_3O_8} + H_2O$$

A considerable portion of the superphosphoric acid is pyrophosphoric acid ($H_4P_2O_7$). Further concentration can be done by producing other polyphosphoric acids. Polyphosphoric acid refers to a series of acids containing more than one atom of phosphorus such as pyrophosphoric acid.

Product Characteristics

Superphosphoric acid made from furnace orthophosphoric acid contains 76% P_2O_5 whereas that made from wet process acid contains 72% P_2O_5. It can be used in conjunction with wet process acid where it serves to sequester the impurities in the acid when it is ammoniated.

It may be used to produce liquid fertilizers of higher analysis than those obtained from wet process phosphoric acid. By using wet process phosphoric acid the highest grade of liquid fertilizer (1 : 3 : 0 ratio) that will not salt out at 0°C is 8-24-0. Using white superphosphoric acid 11-37-0 grade is produced.

Physical and chemical properties of superphosphoric acid are compared with those of furnace grade orthophosphoric acid in Table 3.1.

Superphosphoric acid is highly viscous, colourless transparent liquid. The viscosity of the acid increases as the temperature decreases. However, superphosphoric acid is generally less corrosive than orthophosphoric acid. Stainless steel shows satisfactory resistance to corrosion by superphosphoric acid.

The principal use of superphosphoric acid has been for the manufacture of higher analysis liquid mixed fertilizers that have lower salting out temperatures than those produced with orthophosphoric acid. Superphosphoric acid is added up to 20% by volume to liquid fertilizers made from wet process orthophosphoric acid to act as sequestrant for the impurities of green acid (wet process phosphoric acid). The impurities are complexed by the polyphosphoric acid and held in suspension.

Table 3.1 Properties of orthophosphoric acid and superphosphoric acid

Property	*Orthophosphoric acid*	*Superphosphoric acid*
Specific gravity (27°C)	1.574	1.92
H_3PO_4 content, %	75	105
P_2O_5 content, %	54.3	76
Orthophosphate content, % of total P content	100	49
Viscosity (27°C), centipoises	17	780
Freezing temperature, °C	–18	16
Boiling point, °C	135	315

Selected References

FAI (2001). *Handbook on Fertilizer Technology*. 6th ed., Fertiliser Association of India, New Delhi. pp. 273.

UNIDO/IFDC (Eds) (1998). *Fertilizer Manual*, 3rd ed., Kluwer Academic Publishers, Dordrecht, the Netherlands.

CHAPTER

4

Ammonia Production

Earth's atmosphere contains about 75.5 per cent nitrogen by weight (78.1% by volume). It is estimated that over 90,000 tonnes of nitrogen gas is present in the air over every hectare of earth surface. In the atmosphere, nitrogen exists as diatomic N_2 molecule in which the atoms are joined by a strong triple bond (N≡N) making the N_2 gas inert. The N_2 gas does not react with any element under normal conditions. As such the enormous quantity of nitrogen gas present in the atmosphere cannot be used directly by plants or animals unless converted to a compound. The N_2 gas must combine with at least one other element to be useful. This conversion of atmospheric nitrogen (N_2 gas) to a usable form is referred to as nitrogen fixation.

Early Attempts

The first experimental work involving nitrogen fixation was performed by Joseph Priestley and Henry Cavendish prior to 1800. From their observations, Kristian Birklad and Samuel Hyde, both of Norway, developed the "arc process" of nitric acid production which was analogous to lightning flash in nature. The first commercial plant based on this process was built in Norway in 1905.

In the year 1898, two Germans—Frank and Caro—developed and patented the reaction between calcium carbide and nitrogen to form calcium cyanamide. The process became known as cyanamide process. The first commercial plant utilizing cyanamide process was built in Italy in 1906, the second at Niagara Falls, Ontario, Canada. Other plants later came up in Germany.

Both the above processes were expensive and required heavy electricity consumption. With the advent of synthetic ammonia plants since 1913, arc process and cyanamide process lost their ground and were eventually replaced by ammonia synthesis plants. The first commercial ammonia plant based on Haber-Bosch process was built at BASF complex, Ludwigshafen-Oppau, Germany in 1913. For the invention of the process of ammonia synthesis, Prof. Fritz Haber was awarded Nobel Prize in chemistry in 1918. Later Carl Bosch was also awarded Nobel Prize in chemistry in 1931 for his discovery of the inexpensive iron catalyst.

Because of its many uses, ammonia is one of the most highly-produced inorganic chemicals. There are literally dozens of large-scale ammonia production plants worldwide. It is produced in 80 countries. The worldwide production of ammonia in 2006 was 150 million tonnes. China produced 31.5% of the worldwide production followed by India with 8.8%, Russia with 8.5%, and the United States with 6.8%. Over 85% of the ammonia produced is used for production of fertilizers. Ammonia is also used for the production of plastics, fibres, explosives, and intermediates for dyes and pharmaceuticals.

Haber-Bosch Process

The basic reaction in the Haber-Bosch process is simple combination of nitrogen (N_2) and hydrogen (H_2) gas in 1 : 3 ratio.

$$\underset{\text{(1 volume)}}{N_2} + \underset{\text{(3 volumes)}}{3\,H_2} = \underset{\text{(2 volumes)}}{2\,NH_3} + \text{Heat}$$

Since the reaction is slow, Haber used osmium catalyst, which was very expensive and not easily available. The meticulous efforts of Bosch led to the development of iron catalyst. As the reaction is exothermic and reversible, it follows the Le Chatelier principle of mobile equilibrium. Since the volume of ammonia formed (2 volumes) is less than the volume of original gases (1 vol. N_2 + 3 vol. H_2 = 4 volumes), an increase in the pressure will tend to shift the equilibrium in favour of ammonia leading to higher percentage ammonia at equilibrium. The conversion percentage at 450°C increases from 16 to 69 as the pressure increases from 100 to 1000 atmospheres (one atmosphere = 760 mm of Hg or 1.033 kg cm^{-2} or 14.696 psig). Conversely, an increase in temperature decreases the percentage ammonia at equilibrium. At 400 atm pressure decreasing the temperature from 600°C to 200°C will increase the ammonia percentage from 18 to 93. A low temperature however is not favoured because of slow rate of reaction. In practice, it is necessary to effect a compromise to provide the most favourable temperature and pressure for reasonable rate of reaction and conversion percentage.

Feedstocks

All the nitrogen required in ammonia synthesis is derived from the air. The hydrogen is obtained from various raw materials such as natural gas, oil or coal. These raw materials are called as "feedstock". The feedstocks for ammonia plants may broadly be grouped into three categories, viz. solid, liquid and gaseous. Solid feedstocks are coal, coke, lignite, etc. Liquid feedstocks are naphtha, heavy fuel oil, liquefied petroleum gas and water, whereas gaseous feedstocks include natural gas, refinery gas and coke oven gas. The requirement of feedstock and energy per tonne of ammonia is given in Table 4.1. Detailed description of these three categories of feedstocks has been given in Chapter 2. Choice of a feedstock for the ammonia plant depends mainly on its availability and economics of its processing.

Manufacture of Ammonia

Synthesis of ammonia from any of the feedstocks mentioned above involves several steps which may be broadly grouped into four:

Table 4.1 Energy requirement of different feedstocks for ammonia

Feedstock and Fuel	*Requirements per tonne of ammonia*	*Energy (million kcal)*
Natural gas	860 Nm^3	7.7
Naphtha	0.8–0.9 tonne	8.1
Fuel oil	0.91 tonne	11.5
Coal	3.8 tonne	14.8

Source: FAI (2001)

1. Synthesis gas preparation
2. Shift conversion
3. Gas purification
4. Ammonia synthesis

1. Synthesis Gas Preparation

This step encompasses generation of hydrogen (H_2) gas and suitable introduction of air to get a mixture of N_2 and H_2 in 1 : 3 ratio for ammonia synthesis. The process of generation of H_2 depends largely on the kind of feedstock used as shown in Table 4.2. Some important processes followed for synthesis gas preparation are steam reforming, partial oxidation, coal gasification and electrolysis of water.

Table 4.2 Processes of synthesis gas preparation

S. No.	*Feedstock*	*Process of synthesis gas preparation*
1	Natural gas/Associated gas	Steam reforming or partial oxidation
2	Coke oven gas	Steam reforming or partial oxidation or low temperature separation
3	Naphtha	Steam reforming or partial oxidation
4	Fuel oil/LSHS	Partial oxidation
5	Water	Electrolysis
6	Coal	Partial oxidation

Steam reforming: This is the most popular process which can be adopted to handle gaseous and liquid feedstocks such as natural gas, refinery gas or naphtha. This process operates through endothermic reaction of steam and hydrocarbons over a catalyst, usually nickel, on a refractory base. The two well-known steam reforming processes are the ICI and Haldor Topsoe. In both these processes desulphurization of the hydrocarbon is the first step. Removal of all traces of sulphur from the feedstock is absolutely essential because of its poisonous effect on the catalyst used in the next steps. The sulphur content of naphtha, for example, must be reduced to less than 1 part per million (ppm). This is achieved by converting all sulphur into hydrogen sulphide (H_2S) which is then allowed to react with heavy metals such as iron (Fe_2O_3), molybdenum (Mo), cobalt (Co), sodium ($NaVO_3$) and zinc (ZnO) to precipitate out as metal sulphide.

Steam reforming is usually carried out in two stages by use of primary and secondary reformers. In the primary reformer, the catalyst is placed in tubes which are heated externally whereas the secondary reformer contains a single bed of catalyst. The feed to the primary reformer consists of a mixture of steam and natural gas or naphtha vapour. The tube outlet temperature is between 700 and 830°C and outlet pressure is between 100 and 300 psig. In the secondary reformer, controlled amount of air is introduced to the process gas steam to react with the unreformed methane and to supply the required volume of N_2 gas to make a $N_2 + H_2$ mixture in 1 : 3 ratio. The hot mixture of process gas then flows over a single bed of nickel catalyst at about 950°C in a refractory lined mild steel vessel. In the process gas leaving secondary reformer methane content is merely 0.2%.

The mechanism of reaction during steam reforming is complex and is not fully known. The main reaction however is the endothermic reaction between methane and steam giving rise to H_2 and CO or CO_2.

$$CH_4 + H_2O + \text{Heat} = 3H_2 + CO$$
$$CH_4 + 2H_2O + \text{Heat} = 4H_2 + CO_2$$

The gaseous mixture leaving the secondary reformer is known as synthesis gas. A simplified flow diagram of steam reforming of natural gas or naphtha is shown in Fig. 4.1.

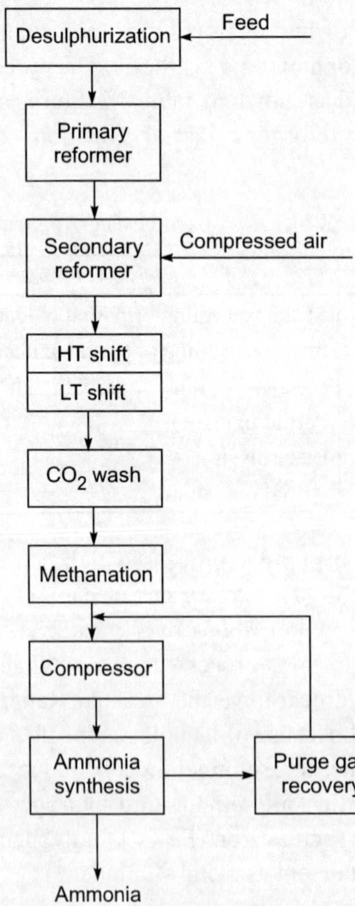

Fig. 4.1 Simplified flow diagram of steam reforming of natural gas or naphtha

Partial oxidation: This process is commonly employed when heavy fuel oil and solid carbonaceous materials are to be used as feedstock. Main advantages of this process are no catalyst requirement, low heat requirement, tolerance of impurities in the feedstock and adaptability to a wide range of hydrocarbon feedstocks.

Two important processes based on partial oxidation of hydrocarbons are those developed by Texaco and Shell. Both these are high pressure non-catalytic processes.

In partial oxidation process, hydrocarbons or heavy molecules are oxidized under pressure to produce H_2 and oxides of carbon. The reaction takes place in a refractory lined combustion chamber at temperature of 1091 to 1482°C and pressure upto 500 psig. The oxidizing agent may be air, oxygen enriched air or pure oxygen. With heavy hydrocarbons steam is added to maintain flow velocity, inhibit cracking during preheating, control carbon formation and reduce high adiabatic flame temperature. Flow diagram for partial oxidation process is shown in Fig. 4.2.

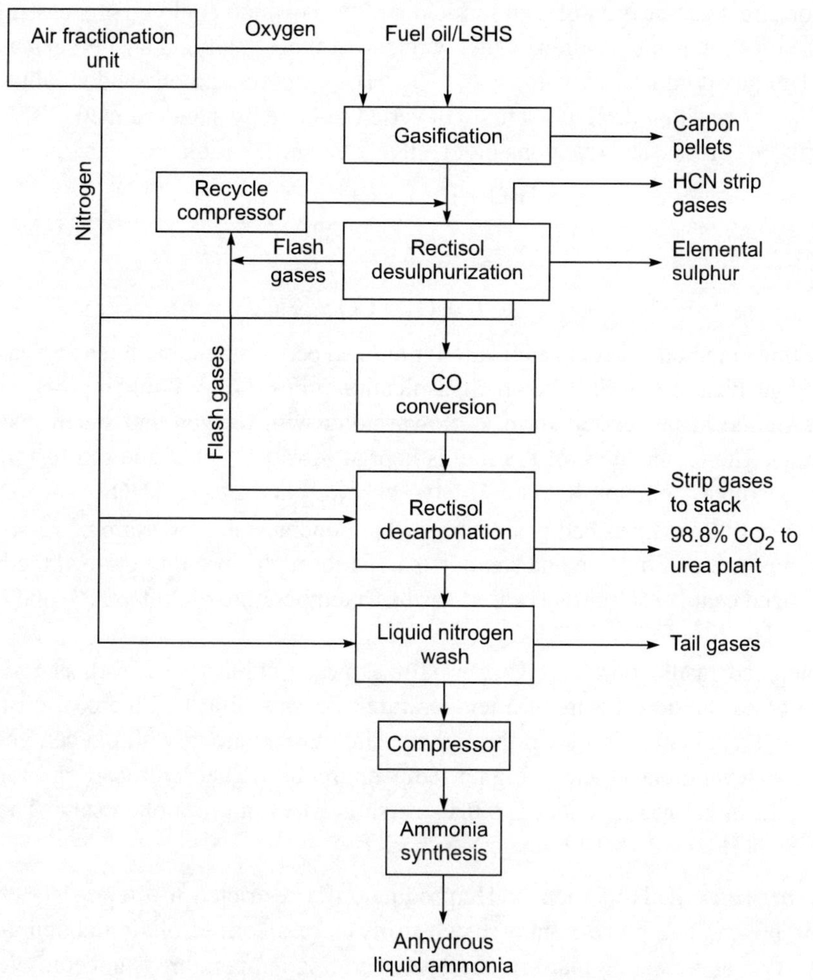

Fig. 4.2 Flow diagram for partial oxidation process

The necessary heat for the process is supplied by the oxidation of methane itself. The main reaction is exothermic combustion of hydrocarbon with limited amount of oxygen to form a flame.

$$CH_4 + \frac{1}{2} O_2 = CO + 2H_2 + Heat$$
$$CH_4 + 2O_2 = CO_2 + 2H_2O + Heat$$
$$CH_4 + H_2O + Heat = CO + 3H_2$$
$$CH_4 + CO_2 + Heat = 2CO + 2H_2$$

Cracking of methane also occurs

$$CH_4 = C + 2H_2$$
$$C + H_2O = CO + H_2$$

Coal gasification: Efforts to produce ammonia on a large-scale using coal as feedstock were first made in Germany in 1915. Since then several processes using oxygen and steam have been developed for direct gasification of coal to yield a raw gas which could be processed for ammonia synthesis. Gasification is the reaction of fuel with one or more gasifying agents such as air, oxygen, steam, etc. The gas produced contains CO, CO_2, H_2, N_2, hydrocarbons and sulphur compounds. Coal can either be gasified directly or first converted to coke by standard methods. Coke consists almost entirely of carbon. The reactions involved in coke gasification are:

$$C + H_2O + Heat = CO + H_2$$
$$C + CO_2 = 2CO$$
$$C + 2H_2 = CH_4 + Heat$$
$$C + O_2 = CO_2 + Heat$$

There are three methods of coal gasification: fluidized bed gasification, fixed bed gasification and entrained bed gasification. In fluidized bed gasification process (e.g. Winkler process), non-caking fuel such as lignite in pulverized form is blown along with oxygen and steam inside a gasifier (gas generator). The temperature of the fuel is kept at 30–50°C below the ash fusion temperature (1000°C). The exit gas contains 40–44% H_2, 16–24% CO_2 and 27–42% CO.

Lurgi process follows fixed bed gasification which operates at a pressure of 25–40 kg cm^{-2}. In this process, a mixture of oxygen and steam is passed through a rotating grate at the bottom of the fixed bed of sized coal (2–50 mm). The exit gas has a temperature of 300–600°C and contains 10% methane, 40% H_2, 20% CO and 30% CO_2.

In entrained bed gasification (e.g. Koppers-Totzek process) reactive as well as non-reactive fuel is gasified at atmospheric pressure at a temperature of about 1500°C. Fine coal dust (90% – 200 mesh) dried to 1% moisture is blown through special burners along with oxygen and little steam in a chamber at several ends. Due to higher temperature, all higher hydrocarbons are cracked and methane content in the gas is reduced to 0.1%. Ash is taken out in molten state. The effluent gas contains 25–30% H_2, 11–15% CO_2 and 52–57% CO.

Electrolysis of water: This method of H_2 production is restricted to the areas having low cost hydroelectric power. The process involves electrolysis of alkaline solution containing potassium hydroxide by two non-reacting plate electrodes. Operating temperature is generally less than 80°C at atmospheric pressure or 90°C at elevated pressure (30 atm).

One major electrolysis plant was constructed in 1958 at Nangal (National Fertilizers Ltd.) but due to non-availability of sufficient power it had to switch over to naphtha in 1998.

2. Shift Conversion

The process gas coming out of steam reforming or partial oxidation or coal gasification contains more than 10% carbon monoxide (CO) which is a poison for ammonia synthesis catalyst. The water gas shift reaction is a convenient and economic way to remove CO while at the same time it increases H_2 concentration. An additional mole of H_2 is produced for each mole of CO shifted. Carbon dioxide that is formed is not only easier to scrub out than CO but also it can be utilized for other products such as urea, dry ice, etc. In this step, the process gas is cooled to 370°C, the steam content is adjusted, if necessary, and then the gas is passed over iron-chrome catalyst. Shift converter operates at a pressure of 14 kg/cm^2 or higher. The reaction is

$$CO + H_2O = CO_2 + H_2 + \text{Heat}$$

In modern plants shift conversion usually takes place in two steps, viz. high temperature shift conversion and low temperature shift conversion. In the first step described above, CO content is reduced to 2–3% by volume whereas in the second step the gas is cooled to about 200°C and then passed over iron-zinc catalyst. This reduces CO content to 0.2–0.5% by volume.

3. Gas Purification

The contaminants which are corrosive to the equipment or poisonous to the ammonia synthesis catalyst such as CO, CO_2, and sulphur compounds must be either completely eliminated or reduced to safe levels. For example, carbon oxides must be reduced to less than 10 ppm. Depending on the technique of H_2 production, sulphur compounds (mainly H_2S) are removed before the raw material is processed to generate H_2. The removal of CO and CO_2 is accomplished in two steps:

Carbon dioxide removal: Two most popular solvents used for CO_2 removal are monoethanolamine (15–20% solution) and hot potassium carbonate. Monoethanolamine (MEA) is most effective solvent which reduces CO_2 to 200–300 ppm and gives 99% pure CO_2 but has high heat requirement and corrosion problem. Hot potassium carbonate absorbs CO_2 by the reaction

$$K_2CO_3 + H_2O + CO_2 = 2KHCO_3$$

Hot potassium carbonate is less effective than MEA but has low heat and cooling requirements. More effective solvents such as sulphinol, fluor, etc. have recently been developed.

Carbon monoxide removal: After shift conversion and CO_2 removal from the process gas, small amount of CO and traces of CO_2 are still present, which must be reduced to less than 10 ppm before synthesis gas is compressed and enters ammonia converter. This is accomplished in the final gas purification step. There are three processes for final gas purification, viz. methanation, copper ammonia solution and liquid nitrogen wash. Methanation involves heating of gas stream to 315°C and passing hot gas through a nickel based catalyst to convert carbon oxides into methane.

$$CO + 3H_2 = CH_4 + H_2O$$
$$CO_2 + 4H_2 = CH_4 + 2H_2O$$

The methane so formed is not removed but steam is removed.

Use of copper-ammonia solution is the oldest method of CO removal. The absorbing solution contains cuprous and cupric ammonium acetate or cuprous and cupric ammonium formate or a mixture of the two.

Liquid nitrogen wash or cryogenic scrubbing gives the complete purification of process gas. It removes not only CO and CO_2 but also argon and methane. The method involves passing the process gas through liquid nitrogen. Since air separation plant is required for producing liquid nitrogen, this method is normally used in conjunction with partial oxidation process of synthesis gas preparation and with use of coke oven or refinery gas as feedstock.

4. Ammonia Synthesis

The purified synthesis gas is compressed to a pressure of 135–300 atm (2000–5500 psig) and passed over a catalyst in the ammonia converter. The ammonia converter is a large, long vertical cylinder often weighing 100 tonnes or more. Modern converters have capacity upto 1000 tonnes ammonia per day. As much as 50 per cent of the space in the converter may be filled with the catalyst. Most popular catalyst is the promoted iron oxide (Fe_2O_3) which is prepared by mixing oxides of iron with small amount of promoters. The promoters are a combination of acidic and alkaline metallic oxides such as oxides of aluminium, potassium, calcium and magnesium. The promoters increase the activity and life of the catalyst.

The catalytic reaction of N_2 and H_2 to form ammonia is exothermic but the reaction will not start below a temperature of 350°C. Once the reaction has begun the gases leaving the converter are used to heat the incoming gases through a system of heat exchangers. Increasing temperature increases reaction rate and ammonia recovery from the effluent gas. But higher temperature favours decomposition of ammonia and reduces the life of catalyst. Satisfactory catalyst performance is obtained at a temperature of 900 to 1050°F (482–566°C). As per Le Chatelier principle, increasing the operating pressure increases ammonia content of the effluent gas.

Though basic process is the same as invented by Haber–Bosch, several improvements have since been made in the selection of operating conditions, choice of catalyst, design of converter internals, method of ammonia recovery and type of recirculation system. In general, ammonia synthesis processes may be grouped into two classes: high pressure and low pressure, on the basis of the pressure at which synthesis gas is introduced into the ammonia converter. Low pressure processes operate in the range of 200–350 atm. Haber-Bosch, USDA, Girdler, Chemico and Mont Cenis proceses fall in this category. High pressure processes operate in the range of 800 to1000 atm. Examples are Claude and UHDA.

Depending upon the process, age of catalyst and flow velocity, 10–30% conversion of synthesis gas to ammonia per pass is achieved. The gas mixture leaving the converter is chilled to 30°F (–1°C) to condense most of ammonia and rest of the gases are mixed with fresh make-up gases, compressed and recycled back to refrigeration exchangers, product separator and converter. The non-reacting gases such as methane, helium and other inert gases will concentrate in the recycle gas. To maintain efficient operation, the concentration of these inert gases is reduced by continuous 'gas purge'.

Physical Characteristics of Ammonia

At normal temperature and pressure, anhydrous ammonia exists as pungent and colourless gas with a liquid density of 637.8 g/L (0°C). At 10°C the gas exerts a pressure of 34.1 kg per square inch (6.45 cm^2). It is stored in steel tanks that have a capacity for 120.4 kg per square inch. Ammonia has critical temperature[\$] of 132.4°C and critical pressure[\$] of 111.5 atm. At 1 atmosphere pressure, boiling temperature of ammonia is –33.3°C and freezing temperature is –77.7 °C.

Gaseous ammonia is lighter than air but on compression and cooling it becomes a liquid which is about 60% as heavy as water. Due to high vapour pressure of anhydrous ammonia, it is transported in pressure containers usually at more than 18 atm pressure. Ammonia is ordinarily not combustible but will burn in the presence of oxygen or air if a source of ignition is present. Ignition temperature of ammonia-air mixture is around 78°C. Flammable and explosive mixtures are formed with air in concentration of 16 to 25% by volume.

Chemical Characteristics

Ammonia contains 82% nitrogen by weight. Ammonia is soluble in water upto 30% by weight, resulting in low pressure solution containing 24% nitrogen that is referred to as *aqua ammonia.* Ammonia is a strong alkali and is used as cleaning agent for floors, walls and metal surfaces.

Because ammonia and aqua ammonia are both basic, the immediate effect of adding them to soils is to raise the pH to above 9 in the zone of application. Ammonia reacts with clay, organic matter and soil moisture. After a few days in well-aerated soil, microorganisms convert ammonia to soluble nitrate which moves the cations such as calcium, potassium and magnesium into growing roots or percolating water. Thus, the final effect of ammonia is to reduce the basicity of the soil and to increase the acidity.

Safety: Gaseous ammonia is extremely irritating to eyes and respiratory system in concentration of 0.05–0.1% by volume. A concentration of 0.17% causes convulsive coughing and may be fatal.

Direct Application of Ammonia as Nitrogen Fertilizer

Anhydrous ammonia is the most economic source of nitrogen and all other nitrogen fertilizers are produced from ammonia. However, special storage, transportation and suitable application equipment are required for anhydrous ammonia which is a liquefied gas. In the developed countries, direct application of anhydrous ammonia to the soil has become a popular farm practice. In India, a beginning is yet to be made.

Industrial Uses of Ammonia

Anhydrous ammonia has many industrial uses as shown in Table 4.3. It is the basic raw material for manufacture of all the nitrogen fertilizers and nitric acid. Besides, ammonia is required in the manufacture of a large number of chemicals and medicines. Explosives like ammonium nitrate,

[\$] Critical temperature is the temperature above which only gaseous phase exits, no matter how great the pressure applied to the system. The pressure that would just suffice to liquefy the gas at the critical temperature is the critical pressure.

synthetic fibres and plastics are produced from ammonia. Ammonia gas is used for refrigeration as also in metallurgy and several other industries.

Table 4.3 Industrial uses of ammonia

Industry	*Use*
Fertilizer	Production of • ammonium sulphate, • ammonium nitrate, • urea, • nitrogen solutions, • ammonium phosphates, • nitrophosphates
Chemicals	Production of • nitric acid which is used for production of nitroglycerine and TNT which is also used as a vasodilator (a substance that dilates blood vessels) and PETN (pentaerythritol nitrate); • sodium hydrogen carbonate (sodium bicarbonate), • sodium carbonate, • hydrogen cyanide (hydrocyanic acid), • hydrazine (used in rocket propulsion systems).
Explosives	Ammonium nitrate
Fibres and plastics	Nylon and other polyamides
Refrigeration	For making ice, large-scale refrigeration plants, air-conditioning units in buildings and plants
Pharmaceuticals	Manufacture of drugs such as sulphonamide which inhibit the growth and multiplication of bacteria that require *p*-aminobenzoic acid (PABA) for the biosynthesis of folic acids; anti-malarials and vitamins such as the B vitamins nicotinamide (niacinamide) and thiamine
Pulp and paper	Ammonium hydrogen sulphite, enables some hardwoods to be used
Mining and metallurgy	Used in nitriding (bright annealing) steel; used in zinc and nickel extraction.
Cleaning	Used as a cleaning agent such as in cloudy ammonia

Selected References

FAI (2001). *Handbook on Fertilizer Technology*. 6th ed., Fertiliser Association of India, New Delhi. pp. 273.

Mordtvet, J.J., Murphy, L.S. and Follett, R.F.H. (1999). *Fertilizer Technology and Application*. Meister Publishing Company, Euclid Ave., Willoughby, OH.

UNIDO/IFDC (Eds.) (1998). *Fertilizer Manual*, 3rd ed., Kluwer Academic Publishers, Dordrecht, the Netherlands.

CHAPTER

5

Nitrogen Fertilizers

Nitrogen fertilizers are produced and consumed in the largest quantity. Historically, the first nitrogen fertilizer was sodium nitrate which was recovered from the natural deposits in Chile. After discovery of ammonia synthesis from nitrogen and hydrogen gases in early twentieth century, a number of solid nitrogen fertilizers were developed, and their production began in many countries.

Classification of Nitrogen Fertilizers

According to the chemical form of nitrogen in the product, the nitrogen fertilizers are divided into six groups: ammonium, nitrate, ammonia and ammonium salts, amide form, liquid and slow release materials.

1. Ammonium-N fertilizers: Ammonium fertilizers are soluble in water. When applied to soils, they are partly adsorbed on the soil colloids. Thus, the NH_4-N is protected from being washed away by run off or by leaching. However, ammonium-N is quickly converted by microbiological nitrification process into nitrate form which is mobile and susceptible to losses. Ammonium fertilizers are acidic in their residual effect in soil. The examples of the ammonium fertilizers are ammonium sulphate, ammonium chloride, anhydrous ammonia and aqua ammonia (Table 5.1).
2. Nitrate-N fertilizers: Nitrogen present in these fertilizers is in nitrate (NO_3^-) form, which is soluble in water and readily absorbed by the plants. Nitrate ions are highly reactive and mobile, and are susceptible to losses due to leaching and under waterlogged conditions by denitrification. Sodium nitrate and calcium nitrate are two fertilizers in this group but they are not popular any more.
3. Ammonium and nitrate-N fertilizers: These fertilizers contain nitrogen in both ammonium and nitrate forms. The nitrate nitrogen is readily available to plants for immediate need, whereas ammonium nitrogen becomes available to plants after a few days, when it is transformed into nitrate. They are soluble in water and suitable for most of the crops and

soils. They are acidic in their residual effect. Ammonium nitrate and calcium ammonium nitrate (CAN) are two fertilizers in this group.

4. Amide fertilizer: This group of fertilizers contains nitrogen in organic compounds as amide, (H_2NCONH_2), which is not directly available to plants as such, but quickly converted by soil microbes to ammoniacal and nitrate forms and then utilized. The only amide fertilizer available presently is urea, which is the most popular N fertilizer in India and other developing countries due to high N content and relatively low unit cost of N.
5. Liquid N fertilizers: Liquid N fertilizers are used in developed countries as their application requires suitable equipment for handling, storage and application. They are commonly known as nitrogen solutions. Nitrogen solutions contain ammonium nitrate, urea and water. They are used in manufacturing multinutrient fertilizers or for the direct application to soil.
6. Slow release fertilizers: These are newly developed fertilizers which release nitrogen in soil very slowly so that it may be available to the plants for a longer period of time. Use of these materials results in better utilization of applied nitrogen by the growing crop plants and reduce losses, thereby reducing the risk of environmental pollution. Ureaform, IBDU, CDU and sulphur coated urea are the fertilizers in this group.

Table 5.1 Commercial nitrogen fertilizers and their N content

S. No.	*Fertilizer*	*Chemical formula*	*% N*
	Ammonium-N sources		
1.	Ammonium sulphate	$(NH_4)_2SO_4$	20.6
2.	Ammonium chloride	NH_4Cl	25
3.	Diammonium phosphate (DAP)	$(NH_4)_2 HPO_4$	18
4.	Anhydrous ammonia	NH_3	82
5.	Aqua ammonia	NH_3 in water	20–25
	Nitrate-N sources		
6.	Sodium nitrate	$NaNO_3$	16.0
7.	Calcium nitrate	$Ca(NO_3)_2$	15.5
	Ammonium + Nitrate-N sources		
8.	Ammonium nitrate	NH_4NO_3	34
9.	Calcium ammonium nitrate (CAN)	$NH_4NO_3 + CaCO_3$	25
	Amide-N sources		
10.	Urea	$CO(NH_2)_2$	46
	Liquid N fertilizers		
11.	Nitrogen solutions (non-pressure)	Urea + ammonium nitrate	28–32
12.	Nitrogen solutions (pressure)	Ammonia + urea + ammonium nitrate	34–38
13.	**Slow release N sources**		
14.	Ureaform		38
15.	Isobutylidine diurea (IBDU)		32.2
16.	Crotonilidine diurea (CDU)		32
17.	Guanyl urea		37
18.	Sulphur coated urea (SCU)		36–40
19.	Neem-coated urea		35–37

Ammonia as the Raw Material

Ammonia is the basic material to provide nitrogen for production of all synthetic nitrogen fertilizers. Anhydrous ammonia is the most concentrated nitrogen fertilizer (82% N) but being a gas at normal atmospheric pressure and temperature it requires special storage, transport and application equipment. This has led to the development of a number of solid nitrogen fertilizers using ammonia as raw material. Some fertilizers such as urea and ammonium nitrate can be manufactured from ammonia and by-product of ammonia plants as no other raw material is required. A variety of single and multinutrient fertilizers as shown in Fig. 5.1 are produced from ammonia.

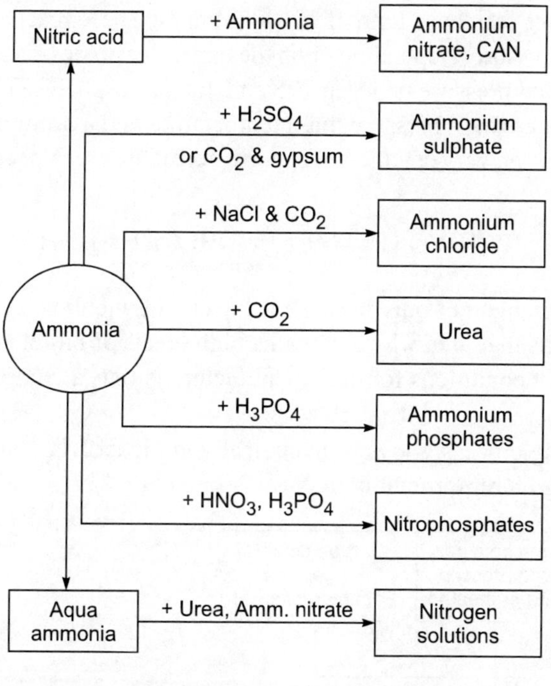

Fig. 5.1 Possibilities of production of nitrogen fertilizers from ammonia

UREA

Urea is the most popular nitrogen fertilizer in India and many other developing countries, mainly because of its high nitrogen content (46%). In 2006–07, about 82% of the total nitrogen fertilizer manufactured in India was in the form of urea.

Raw Materials

Principal raw materials required to produce urea are ammonia and carbon dioxide.

Method of Manufacture

Urea is produced from ammonia and carbon dioxide in two equilibrium reactions:

$$2NH_3 + CO_2 = \underset{\text{Ammonium carbamate}}{NH_2COONH_4} + \text{Heat}$$

$$\underset{\text{Ammonium carbamate}}{NH_2COONH_4} = \underset{\text{Urea}}{NH_2CONH_2} + H_2O$$

The first reaction leading to formation of ammonium carbamate is essentially instantaneous. But the second reaction which results in synthesis of urea by dehydration of ammonium carbamate is relatively slow and requires residence time of 20–30 minutes. These reactions are accomplished in an autoclave at elevated pressure. Depending upon the particular process, temperature is maintained between 170 and 200°C and pressure between 175 and 400 atmospheres. The operating conditions most suited for high yield are high pressure, high temperature and a minimum excess of ammonia. These conditions are, however, very conducive for corrosion of urea reactor equipment and formation of biuret by the reaction:

$$\underset{\text{Urea}}{2NH_2CONH_2} = \underset{\text{Biuret}}{NH_2CONHCONH_2} + NH_3$$

This reaction is undesirable, not only because it lowers the yield of urea, but because biuret is toxic to plants. This means that urea which contains high levels of biuret is unsuitable for use as a fertilizer. Selection of best conditions for urea manufacture is thus a compromise between several competing factors.

The urea manufacturing process, shown schematically in Fig. 5.2, is designed to maximize these reactions while inhibiting biuret formation:

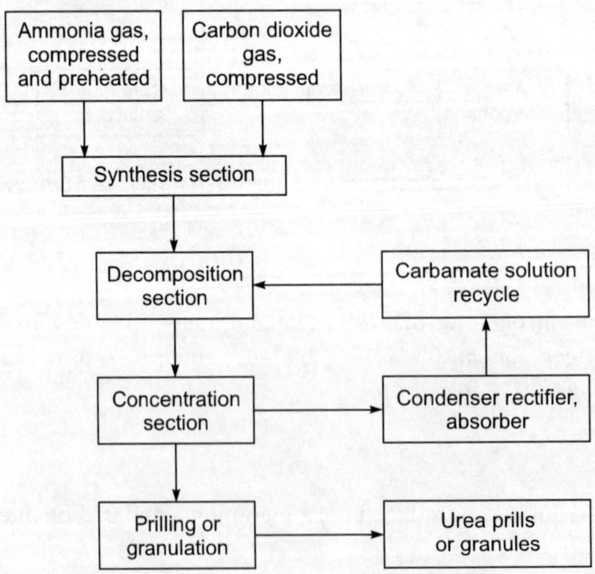

Fig. 5.2 A typical urea production process

There are four main sections in a urea plant, viz. reactor (autoclave), carbamate decomposer, prilling and recycling of unreacted ammonia and carbon dioxide. Major difference among various commercial processes of urea manufacture lies in the method of treating the off gases (ammonia and carbon dioxide). Depending on the degree of recycling of these gases, a process is called once through, partial recycle or total recycle process.

Urea manufacture proceeds in *four* major steps:

Step 1 – Synthesis

A mixture of compressed CO_2 and ammonia is reacted to form ammonium carbamate. This is an exothermic reaction, and heat is recovered by a boiler which produces steam. Typical operating conditions include temperatures from 180 to 200°C, pressures from 140 to 250 atmospheres, $NH_3 : CO_2$ molar ratios from 3 : 1 to 4 : 1, and a retention time of 20 to 30 minutes. The carbamate is then dehydrated to yield 70 to 77 per cent aqueous urea solution. The solution then goes for purification.

Step 2 – Purification

The major impurities in the mixture at this stage are water from the urea production reaction and unconsumed reactants (ammonia, carbon dioxide and ammonium carbamate). The unconsumed reactants are removed in three stages. Firstly, the pressure is reduced from 240 to 17 barg (1 atmosphere = 14.50378 PSI = 1.01325 bar = 0 barg) and the solution is heated, which causes the ammonium carbamate to decompose to ammonia and carbon dioxide:

$$\underset{\text{Ammonium carbamate}}{NH_2COONH_4} = 2NH_3 + CO_2$$

At the same time, some of the ammonia and carbon dioxide flash off. The pressure is then reduced to 2.0 barg and finally to –0.35 barg, with more ammonia and carbon dioxide being lost at each stage. By the time the mixture is at –0.35 barg, a solution of urea dissolved in water and free of other impurities remains. At each stage the unconsumed reactants are absorbed into a water solution which is recycled to the secondary reactor. The excess ammonia is purified and used as feedstock to the primary reactor.

Step 3 – Concentration

There are three methods of concentrating the urea solution: vacuum concentration, crystallization, and atmospheric evaporation. The method chosen depends upon the level of biuret ($NH_2CONHCONH_2$) impurity permissible in the end product. Aqueous urea solution begins to decompose at 60°C (140°F) to biuret and ammonia. The most common method of solution concentration is evaporation. About 75% of the urea solution is heated under vacuum, which evaporates off some of the water, increasing the urea concentration from 68% w/w to 80% w/w. At this stage some urea crystals also form. The solution is then heated from 80 to 110°C to redissolve these crystals prior to evaporation. In the evaporation stage, molten urea (99% w/w) is produced at 140°C. The remaining 25% of the 68% w/w urea solution is processed under vacuum at 135°C in a two series evaporator-separator arrangement.

Step 4 – Granulation

The concentration process furnishes urea "melt" for solids formation. Urea solids are produced from the urea melt by two basic methods: prilling and granulation. Prilling is a process by which solid particles are produced from molten urea. Molten urea is sprayed from the top of a prill tower. As the droplets fall through a countercurrent airflow, they cool and solidify into nearly spherical particles.

There are two types of prill towers: fluidized bed and non-fluidized bed. The major difference is that a separate solids cooling operation may be required to produce agricultural grade prills in a non-fluidized bed prill tower.

Granulation is used more frequently than prilling in producing solid urea for fertilizer. Granular urea is generally stronger than prilled urea, both in crushing strength and abrasion resistance. There are two granulation methods: drum granulation and pan granulation. In drum granulation, solids are built up in layers on seed granules placed in a rotating drum granulator/cooler approximately 4.3 metres (14 feet) in diameter. Pan granulators also form the product in a layering process, but different equipment is used. The solids cooling operation is generally accomplished during solids formation, but for pan granulation processes and for some agricultural grade prills, some supplementary cooling is provided by auxiliary rotary drums.

The solids screening operation removes offsize product from solid urea. The offsize material may be returned to the process in the solid phase or be redissolved in water and returned to the solution concentration process.

The final product is cooled in air, weighed and conveyed to bulk storage ready for packaging or sale. This process of urea production is used to discourage two undesirable reactions — urea hydrolysis and biuret formation.

Product Characteristics

Physical characteristics: Urea prills or granules are white in colour having specific gravity of 1.335 (20°C). It is hygroscopic having critical relative humidity (CRH) of 80.8, 75.8 and 72.5 at 20, 25 and 30°C, respectively. It absorbs moisture from air quickly and becomes sticky, forms lumps or cake on drying. Its storage requires special attention. It mixes well with other fertilizers but the mixtures must be used immediately to avoid the problem of stickiness and caking.

Chemical characteristics: Fertilizer grade urea contains 46.0% nitrogen as against 46.6% in technical grade pure urea. It is highly soluble in water to the extent of 100 g/100 g at 20°C. Urea hydrolyses quickly in contact with water releasing ammonia gas. In soils the urea hydrolysis is mediated by urease enzyme to yield ammonium carbonate which undergoes nitrification to produce nitrate. Under tropical conditions most of the urea is hydrolyzed in a moist soil within 3–4 days.

$$\underset{\text{Urea}}{NH_2CONH_2} + \underset{\text{Water}}{2H_2O} = \underset{\text{Ammonium carbonate}}{(NH_4)_2CO_3}$$

$$(NH_4)_2CO_3 = 2NH_3 + CO_2 + H_2O$$

AMMONIUM SULPHATE

Ammonium sulphate is one of the oldest fertilizers. It is manufactured either as principal product or by-product.

Raw Materials

The raw materials required for manufacture of ammonium sulphate depend on the process to be followed. They are:

- Sulphuric acid for its recovery from coke oven gas
- Ammonia and sulphuric acid for direct neutralization process
- Ammonia, gypsum and carbon dioxide for production by gypsum process

Methods of Manufacture

Ammonium sulphate is manufactured in several ways such as recovery from coke ovens, direct neutralization process, gypsum process and by-product of caprolactum.

Recovery from Coke Ovens

Ammonium sulphate is produced as by-product of steel industry from the ammonia recovered from coke oven gas. Bituminous coal contains 1–2% nitrogen, of which only 15–20% is recovered as ammonia after coking. Thus, about 2.5–3.0 kg of ammonia is obtained per tonne of coal or 7–12 kg of ammonium sulphate is produced per tonne of coal depending on the type of coal and coking conditions. This amounts to 14–18 kg of ammonium sulphate corresponding to each tonne of steel produced.

Coke oven gas contains about 1% ammonia by volume. This gas is cooled and passed through saturators of special design containing weak sulphuric acid. Ammonium sulphate crystals are formed in the saturator, which are recovered, centrifuged, washed and dried.

Direct Neutralization Process

In this process, ammonia gas is directly neutralized with sulphuric acid in saturator-evaporator under vacuum or at atmospheric pressure to produce ammonium sulphate

$$2NH_3 + H_2SO_4 = (NH_4)_2SO_4 + \text{Heat}$$

The reaction is exothermic and the heat of reaction is removed by evaporating water from the feed acid or added to the system for temperature control. The slurry that is formed is pumped to crystallizer unit. The crystal growth and size can be controlled by temperature, pH, length of time the crystals remain in saturator and by the liquor recycle rate. For obtaining white crystals, a pH less than 7 is maintained. The crystals are taken off from the bottom of the crystallizer, centrifuged and the mother liquor is recycled to the saturator. Ammonium sulphate liquor is corrosive; hence corrosion inhibitors such as phosphoric acid or arsenic compounds are added in some cases.

Amorphous ammonium sulphate is produced by reacting ammonia gas and sulphuric acid in spray towers. The heat of reaction removes all the water present and a dry, fine product is continuously collected from the base of the tower. This product is suitable for making mixed and granular fertilizers.

Granular ammonium sulphate is made by combining ammonia and sulphuric acid in a pipe reactor, to produce an ammonium sulphate solution. The solution is sprayed into a granulating drum where the fertilizer granules are formed.

Gypsum Process

Gypsum process of ammonium sulphate production is also known as Merseberg process. In this process, sulphuric acid is not needed and lime ($CaCO_3$) is obtained as by-product. The process is best suited to the areas where naturally occurring gypsum or by-product gypsum from phosphoric acid plant is available.

In this process, carbon dioxide is absorbed in ammonia solution to form ammonium carbonate. Ammonium carbonate is then reacted with gypsum (calcium sulphate) to produce ammonium sulphate (Fig. 5.3).

$$NH_3 + H_2O = NH_4OH$$
$$2NH_4OH + CO_2 = (NH_4)_2CO_3 + H_2O$$
$$(NH_4)_2CO_3 + CaSO_4.2H_2O = (NH_4)_2SO_4 + CaCO_3 + 2H_2O$$

Lime ($CaCO_3$) is removed by filtration. Ammonium sulphate solution is evaporated, crystallized, centrifuged and dried. At present, no plant in India is producing ammonium sulphate by this process.

By-product Ammonium Sulphate from Caprolactum

Ammonium sulphate solution is formed during manufacture of caprolactum (the starting material for Nylon-6). The waste liquor containing 35% ammonium sulphate is concentrated and sent to crystallizer unit from where ammonium sulphate crystals are recovered by centrifugation and then dried.

Product Characteristics

Physical characteristics: Ammonium sulphate is a white or light gray crystalline salt. It is also available as amorphous form and in some developed countries it is offered as granular product. Its specific gravity is 1.769 and bulk density is 962 kg/m^3. It is slightly hygroscopic with CRH of 81. It is free flowing and does not pose any problem in handling and storage. However, it is susceptible to caking under high humidity. It can be mixed well with other fertilizers to produce multinutrient fertilizers but the mixture will absorb moisture and will not remain free flowing for long period.

Chemical characteristics: Ammonium sulphate contains 21% ammoniacal-N and 24% sulphate-sulphur. The product obtained from direct neutralization process contains some free acid whereas that obtained from gypsum process has some impurity of gypsum (0.045%). It may contain very small amount of arsenic (0.01%).

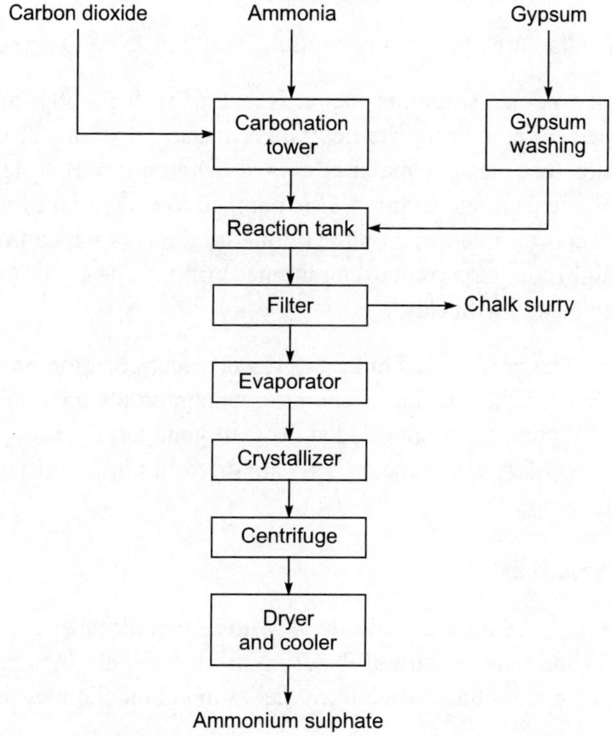

Fig. 5.3 Simplified flow diagram for Merseberg process of ammonium sulphate production

Ammonium sulphate is soluble in water to the extent of 70.6 g per 100 g at 0°C and 103 g per 100 g at 100°C. Ammonium sulphate tends to develop an acid residue in the soil. Its continuous use for a long time will make the soil more acidic.

AMMONIUM NITRATE

Ammonium nitrate is not used as a fertilizer in India but it is a very popular fertilizer in USA and Europe.

Raw Materials

Ammonia and nitric acid are required for manufacture of ammonium nitrate but nitric acid is produced from ammonia.

Method of Manufacture

Ammonium nitrate is produced by several processes that differ slightly but utilize the same basic product, ammonia, to produce nitric acid, which in turn is neutralized by ammonia to form ammonium nitrate. The exothermic reaction is

$$HNO_3 + NH_3 = NH_4NO_3 + \text{Heat}$$

Two methods of manufacture are pressure neutralization process and crystallization process.

Pressure neutralization process: Ammonia vapour is reacted with 50–60% nitric acid in a stainless neutralizer accompanied by agitation. The neutralizer usually operates at 4–5 atm pressure and 170–180°C temperature. The heat of reaction causes evaporation of water and is sufficient to produce 80–85% ammonium nitrate solution. It is pumped to a vacuum evaporator to concentrate further to 94–99%. The concentrated ammonium nitrate solution is sprayed from the top of a prilling tower (60–75 metres high) countercurrent to conditioned airflow. The dry prills (0.2–0.3% moisture) are collected, dried and coated with clay.

Crystallization process: The process is similar to pressure neutralization process upto the point of evaporation. The liquor from the vacuum evaporator containing 80–85% solid is fed to a vacuum crystallizer where crystal growth is controlled at 40°C to yield large grains. The slurry containing about 40% crystals is centrifuged and the crystals are dried in a rotary drier at 85°C. The mother liquor is returned to the evaporator.

Product Characteristics

Physical characteristics: Ammonium nitrate is white crystalline or granular product having specific gravity of 1.66 and bulk density of 998.7 kg/m^3. It is highly hygroscopic having CRH of 58.5 at 30°C. Crystalline ammonium nitrate forms cakes in humid climates. It is therefore stored in air-conditioned silos.

The granular product is free flowing but it should be handled carefully because it is explosive when mixed with combustible materials or exposed to high pressure. When exposed to sparks or open flame ammonium nitrate is a hazardous material.

Chemical characteristics: Pure ammonium nitrate contains about 35% nitrogen. One-half of the nitrogen is in ammoniacal form and the other half in the nitrate form. Ammonium nitrate is soluble in water to the extent of 118.2 g per 100 g of water.

Ammonium nitrate tends to develop an acid residue in the soil. Its continuous use for a long time will make the soil more acidic.

CALCIUM AMMONIUM NITRATE

In order to reduce the explosive hazards of ammonium nitrate, inert material like lime is added so that the fertilizer could be safely handled and used. The product is then marketed as calcium ammonium nitrate (CAN).

Raw Materials

The raw materials required to produce CAN are ammonia, nitric acid, limestone or dolomite and soapstone (coating agent). Nitric acid is produced from ammonia.

Method of Manufacture

Ammonium nitrate solution is prepared by reacting preheated ammonia with nitric acid in a neutralizer. Ammonia is preheated to 85°C by the vapours from neutralizer which also preheats nitric acid to about 65°C. Ammonium nitrate liquor of 60–65% concentration coming out of neutralizer is further concentrated to 92–94% in a vacuum evaporator heated with steam. Concentrated ammonium nitrate is pumped and sprayed over limestone powder in a rotary drum granulator. The granules are dried in a rotary drier and screened. The granules of proper size are cooled in rotary cooler by air and coated with soapstone dust in a coating drum.

Product Characteristics

Physical characteristics: CAN is a greyish granular product and free flowing. It is however hygroscopic (CRH of ammonium nitrate is 58.5 at 30°C) and presents problems in storage. It should be stored in air-conditioned silos.

Chemical characteristics: CAN contains 25% nitrogen in the form of ammonium nitrate. One-half of the nitrogen is in ammoniacal form and the other half in the nitrate form. Nitrogen component of CAN is water soluble but limestone part is insoluble. It also contains sufficient lime to neutralize the acidity produced by ammonium nitrate component. CAN is therefore considered as neutral fertilizer as it does not leave acidic or alkaline residue in the soil.

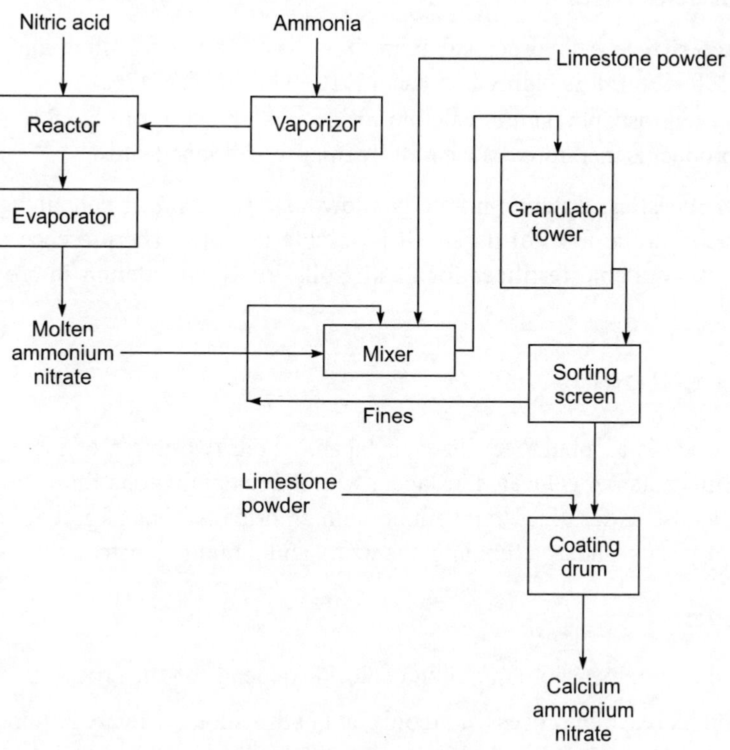

Fig. 5.4 Simplified flow diagram for calcium ammonium nitrate production

CALCIUM NITRATE

Calcium nitrate is not produced in India but is imported for use in industry and agriculture.

Raw Materials

The raw materials required for manufacture of calcium nitrate are limestone and nitric acid.

Method of Manufacture

There are two methods of manufacture of calcium nitrate.

Direct neutralization: Ground limestone is reacted with 50–60% nitric acid in a reactor. The slurry is filtered to obtain 40–50% solution to which small amount of ammonia is added to raise the pH to 4.0. The solution is then pumped to three-stage evaporator, followed by granulation or prilling and coating with anticaking agent.

By-product of nitrophosphate process: Calcium nitrate is a by-product of nitrophosphate plant. However, it is in the form of calcium nitrate tetrahydrate. This form of calcium nitrate carrying 4 molecules of water melts when heated. The melt contains 70% calcium nitrate which is further concentrated followed by granulation or prilling.

Product Characteristics

Physical characteristics: Anhydrous calcium nitrate is a white crystalline material having bulk density 1050–1100 kg/m^3. It is highly hygroscopic (CRH at 30°C is 47%).

Due to highly hygroscopic nature, calcium nitrate absorbs moisture quickly from the air. The storage of this product is therefore done carefully in air-conditioned silos.

Chemical characteristics: Calcium nitrate is a low analysis fertilizer containing 15.5% nitrogen which is essentially in the form of nitrate. It is soluble in water. The presence of 18.8% soluble calcium makes it a suitable fertilizer for alkali soils in which calcium improves the physical properties of the soil.

AMMONIUM CHLORIDE

Ammonium chloride is a popular fertilizer in Japan. About two-thirds of the world capacity for manufacture of this material is located in Japan, with the remaining one-third situated in India and China. Besides its use as fertilizer, pure ammonium chloride (technical grade) is also used in the manufacture of dry cells, and as a flux in galvanizing and tinning industry.

Raw Materials

The raw material requirement for ammonium chloride depends on the process of manufacture.

- For the direct reaction process, ammonia and hydrochloric acid are required.
- For the dual salt process or modified Solvay process, ammonia, carbon dioxide and common salt (sodium chloride) are needed.

Methods of Manufacture

There are two processes for manufacture of ammonium chloride.

Direct reaction process: Ammonia and hydrochloric acid are directly reacted to produce ammonium chloride.

$$NH_3 + HCl = NH_4Cl$$

Ammonia gas is bubbled into hydrochloric acid solution in a reactor. The reaction is controlled by addition of water. The slurry from the saturator at 80% solid concentration is centrifuged and the crystals are washed with water and dried. This process is however not widely used.

Dual-salt process: This is also known as modified Solvay process which is most widely used process for manufacture of soda ash and ammonium chloride. The raw materials required in this process are common salt (sodium chloride), ammonia and carbon dioxide. In this process, 30% solution of ammonia is treated with carbon dioxide in a carbonating tower to form ammonium carbonate.

$$2NH_3 + H_2O + CO_2 = (NH_4)_2CO_3$$

Further carbonation produces ammonium bicarbonate.

$$(NH_4)_2CO_3 + CO_2 = 2NH_4HCO_3$$

Ammonium bicarbonate as it is formed reacts with sodium chloride to give sodium bicarbonate and ammonium chloride.

$$NH_4HCO_3 + NaCl = NaHCO_3 + NH_4Cl$$

The sodium bicarbonate precipitates out and it is separated by filtration, washed and dried to produce soda ash (washing soda).

The ammonium chloride solution, after separation of sodium bicarbonate, is ammoniated, cooled below 215°C and salted out adding washed sodium chloride. The crystals of ammonium chloride formed are centrifuged, washed and dried in rotary drier. The sodium chloride content is reduced by repeated washing. The mother liquor after removal of sodium chloride is re-ammoniated and recycled to carbonating tower.

Product Characteristics

Physical characteristics: Ammonium chloride is a white crystalline product having specific gravity 1.53, bulk density 4.5–5.0 kg/m^3. It is free flowing and non-abrasive. It is however hygroscopic (CRH at 30°C is 77%). It is susceptible to caking at high humidity. Normally, it does not pose any problem in handling and storage.

Chemical characteristics: Fertilizer-grade ammonium chloride contains 25% nitrogen. It is soluble in water.

It is as acid forming as ammonium sulphate per unit of N, and this effect is undesirable in acidic soil, especially if liming costs are excessive.

SODIUM NITRATE

Sodium nitrate is also called soda niter or Chilean saltpeter. It is found naturally in large deposits in arid regions of Chile, Peru, Argentina, and Bolivia as caliche, a crude, impure nitrate rock or gravel. Natural deposits are the major source of sodium nitrate. The world's largest natural deposits of caliche ore were in the Atacama desert of Chile, and many deposits were mined over a century, until the 1940s, when its value declined dramatically in the first decade of the twentieth century as a consequence of widespread use of Haber-Bosch process of ammonia synthesis.

Chile still has the largest reserves of caliche, with active mines in such locations as Pedro de Valdivia, Maria Elena and Pampa Blanca. Sodium nitrate, potassium nitrate, sodium sulphate and iodine are all obtained by the processing of caliche.

Sodium nitrate is also obtained in small amounts as a by-product of chlorine production by the nitrosyl chloride process, in which sodium chloride (common salt) is reacted with nitric acid.

Sodium nitrate was used extensively as a fertilizer and a raw material for the manufacture of gunpowder in the late 19th century. Sodium nitrate is used as an ingredient in fertilizers, pyrotechnics, as a food preservative, and as a solid rocket propellant, as well as in glass and pottery enamels. It was formerly an important raw material for the production of nitric acid. Sodium nitrate is not a popular fertilizer any more.

NITROGEN SOLUTIONS

Nitrogen solutions are used almost exclusively as nitrogen fertilizer materials. There is little or no industrial market. Nitrogen solution fertilizer use is largely confined to countries that have highly developed agricultural economies and very little consumption occurs in the developing regions of the world. Although nitrogen solutions account for only about 7% of the world nitrogen fertilizer market, they account for 35–40% of the market in the United States and 10–15% in Western Europe. Urea-ammonium nitrate (UAN) solutions, containing 28–32% nitrogen, account for the bulk of the solutions market.

Nitrogen solutions tend to be somewhat more expensive per unit of nitrogen than other forms of nitrogen fertilizer, but they have the advantage that they can be applied via irrigation at the exact time during the growing cycle that the application will have maximum impact. They can also incorporate pesticides.

Nitrogen solutions are made by combining various amounts of anhydrous ammonia, ammonium nitrate, urea and water. They are used in manufacturing multinutrient fertilizers or for the direct application to soil. Nitrogen solutions provide economical source of nitrogen, compared to anhydrous ammonia which requires special handling, storage and application machinery.

Nitrogen solutions can be grouped into two categories: pressure solutions and non-pressure solution. Pressure solutions contain ammonia with ammonium nitrate with or without urea and they have higher nitrogen content than non-pressure solutions. Non-pressure solutions do not contain ammonia as they are made by combining various amounts of ammonium nitrate, urea and water. One advantage of non-pressure nitrogen solutions is that they are widely used as diluents for pesticides. They serve as diluents and at the same time supply nitrogen for the crop.

Method of Manufacture

Aqua ammonia (20–25% N) is the simplest nitrogen solution which is made by forcing compressed ammonia into a closed container of water. It has a pressure of less than 10 psi and usually is composed of 25 to 29% ammonia by weight.

The ammonium nitrate liquor prior to being pumped to the top of prilling tower is taken out and may be processed into nitrogen solution for direct application or may be combined with urea to form ammonium nitrate-urea solutions. Use of liquor before prilling step begins is cost-effective as it eliminates six steps used in the processing of solid nitrogen fertilizers. These steps are evaporating, prilling, drying, cooling, coating and bagging.

Product Characteristics

Physical characteristics: Specific gravity of nitrogen solutions varies from 1.18 to 1.38 and weight per gallon varies from 4.5 to 5.2 kg/gallon (3.8 L). Pressure nitrogen solutions have vapour pressure of 1 psig (70 g/cm^2) at 40°C. They need to be stored in pressure tanks. Non-pressure solutions do not have this characteristic and can be stored in ordinary tanks. The salting out temperature (at which crystals begin to form) varies widely from –10°C to 6.8°C depending on the composition.

Chemical characteristics: Ammonium nitrate solutions for direct application contain 19–21% nitrogen. Nitrogen solutions made of ammonium nitrate and urea contain 30–32% nitrogen. The grade of the nitrogen solutions is reported using a standard code which gives % N with content of free ammonia, ammonium nitrate and urea in the order. For example, in nitrogen solution code 370 (17–67–0), 370 refers to 37.0% N and figures in the parenthesis indicate % of free ammonia, ammonium nitrate and urea in that order.

There are three physical and chemical characteristics of nitrogen solutions that influence their use as fertilizers: vapour pressure, salting out temperature and corrosiveness. For safety considerations, nitrogen solutions are stored in mild steel tanks.

The composition and properties of three principal non-pressure solutions are shown in Table 5.2.

Table 5.2 Composition and properties of three principal non-pressure solutions

Particulars	*Grade*		
	28	30	32
Composition by weight (%)			
Ammonium nitrate	40.1	42.2	43.3
Urea	30.0	32.7	35.4
Water	29.9	25.1	20.3
Specific gravity at 15.6°C	1.283	1.303	1.32
Salting out temperature, °C	–18	–10	–2

Slow Release Nitrogen Fertilizers

All nitrogen fertilizers are soluble in water and susceptible to loss through leaching, ammonia volatilization and denitrification. The loss of nitrogen begins the moment nitrogen fertilizer is applied to soil. Leaching loss of N is severe from light textured permeable soils, particularly in case of nitrate-sources of N, whereas ammonia volatilization is pronounced from alkaline soils and flooded rice fields in case of urea and ammonium sources of N. Denitrification occurs in poorly drained soils but it is most severe in flooded rice soils and results in evolution of nitrogen oxides (N_2O and NO) from the soils. Due to continuous loss of N from soils, the recovery efficiency of applied nitrogen fertilizer is very poor, seldom exceeding 40% for rice and 50–60% for upland crops like maize, wheat, etc. The amount of nitrogen not utilized by the crop does not stay in the soil for use by the next crop. Rather, it is lost through various pathways indicated above. The nitrogen lost through leaching causes nitrate enrichment of underground water spoiling its quality as potable water. Ammonia and nitrogen oxides escaping to air pollute the upper atmosphere.

Thus, development of nitrogen fertilizers with greater efficiency is desirable. For this, there is need to develop a nitrogen fertilizer which releases N slowly matching the plant need for N during growth period. The plant N requirement is dynamic as it changes with crop growth. It is low at crop establishment, increases gradually with growth and is at the peak before flowering (end of vegetative growth period). The ideal product would be one that releases N in accordance with plant needs throughout the growing season. The slow release materials would also reduce the need for repeated application of conventional soluble products.

In view of enormous amount of nitrogen fertilizers used, huge economic loss and environmental concerns, interest in development of slow release nitrogen fertilizers has grown remarkable and a number of slow release materials have been developed. Most of the materials that have been developed for slow release of N can be grouped as follows:

1. Materials of low solubility that dissolve slowly in soils
2. Nitrification inhibitors

Materials of low solubility: This group includes the substances that are slightly soluble in water and must undergo chemical or microbial decomposition to release the available N inside the granule and those N fertilizers coated with insoluble substances. The rate of release of N from them is related to their water solubility and the rate of chemical/microbiological action. The products of low solubility materials are ureaform, CDU and IBDU whereas best known coated material is sulphur-coated urea (SCU).

Ureaform is a generic name which refers to a mixture of compounds which are formed by the reaction of urea with formaldehyde. They consist mainly of methylene urea polymers varying in chain length. Small molecules are water soluble whereas large molecules are sparingly soluble. Ureaforms are white, odourless solids which typically contain 38% N about 30% of which is soluble in cold water (25°C). About 40% of the N insoluble in cold water should be soluble in hot water for good crop response. Because of high cost, ureaform is presently used in non-farm markets for turf grass, ornamental plants, horticulture, greenhouse crops, landscaping and forest seedlings.

CDU is abbreviation of crotonylidene diurea which is produced by the condensation reaction between urea and acetaldehyde or crotonaldehyde (two moles of acetaldehyde condensed to one mole of crotonaldehyde). The product contains 32% N which is sparingly soluble in water.

IBDU is abbreviation of isobutylidenediurea which is the major reaction product of urea and isobutyraldehyde. Pure IBDU is a white crystalline material containing 32% N. However, fertilizer grade IBDU is granular non-hygroscopic material that contains 30% N.

Guanylurea is produced as nitrate, phosphate and sulphate salts by direct acidulation of calcium cyanamide. Guanyl urea nitrate contains 37% N while phosphate salt contains 27.8% N. It is produced in small quantity in Japan.

Sulphur-coated urea (SCU) is controlled release N fertilizer consisting of a sulphur shell around each urea particle. It contains 36–38% N in the form of urea. The thickness of sulphur coating around urea particles determines the release rate of N from SCU. The sulphur coating must be oxidized by soil microorganisms before the urea dissolves in the soil and hydrolyses into ammonia. SCU has the greatest potential for use in multiple applications of soluble N sources that are needed during the growing season, particularly on sandy soils under high rainfall or irrigation. It is used in sugar cane, fruits, turf, and rice. Sulphur content of SCU is an additional advantage.

Polymer-coated urea is reported to have better nutrient release pattern than other-coated materials. The rate of nutrient release from polymer-coated urea is not affected by soil environment like acidity, temperature, microbial activity, etc.

Nitrification inhibitors: Nitrification inhibitors are substances which are added to urea and ammoniacal-N materials to retard the process of nitrification by specifically inhibiting the activity of *Nitrosomonas* spp. which accomplishes the oxidation of ammonia into nitrite. This conserves the ammonium-N in the soil and may reduce the loss of N through leaching and denitrification but may enhance the loss by ammonia volatilization under certain environment.

A number of organic compounds have been reported to act as nitrification inhibitor but two generally accepted compounds are N-SERVE or Nitrapyrin (2-chlor-6(trichloromethyl)) pyridine and AM which is a substituted pyrimidine (2-amino-4-chloro-6-methylpyrimidine).

Dicyandiamide is another compound which has been used in Germany and Japan both as nitrification inhibitor and as a slow release nitrogen source. In Japan, a mixed product of ureaform and 10% dicyandiamide is produced. In Germany, a fertilizer composed of urea and DCD in 4 : 1 ratio is commercially available. Urea containing 1.4% DCD is also marketed in North America. The effectiveness of nitrification inhibitors under tropical field conditions is much reduced due to rapid decomposition by soil microorganisms.

Neem cake-coated urea has been developed as slow release fertilizer by blending finely powdered *neem* (*Azadirachta indica*) seed cake with urea granules using bitumen dissolved in kerosene as sticker. The amount of *neem* cake varies from 20 to 30%. It is believed that small amount of nimbidin present in the neem seed cake after oil extraction acts as nitrification inhibitor and slows down the process of nitrate formation from ammonia released from urea hydrolysis. The product is recommended mainly for flooded rice in India and other countries of South Asia.

Selected References

FAI (2001). *Handbook on Fertilizer Technology*. 6th ed., Fertiliser Association of India, New Delhi. pp. 273.

Havlin, J.L., Beaton, J.D., Tisdale, S.L. and Nelson, W.L. (2005). *Soil Fertility and Fertilizers*, 7th ed., Prentice-Hall of India, New Delhi.

Mordtvet, J.J., Murphy, L.S. and Follett, R.F.H. (1999). *Fertilizer Technology and Application*. Meister Publishing Company, Euclid Ave., Willoughby, OH.

UNIDO/IFDC (Eds.) (1998). *Fertilizer Manual*, 3rd ed., Kluwer Academic Publishers, Dordrecht, the Netherlands.

CHAPTER

6

Phosphate Fertilizers

"We may be able to substitute nuclear power for coal power, and plastics for wood, and yeast for meat, and friendship for isolation but for phosphorus, there is neither substitute nor replacement."

–Isaac Asimov

The original source of phosphorus in the early period of manufacture of phosphate fertilizers was bones but the supply was soon exhausted. Today, essentially all phosphorus in various industrial products such as fertilizers, detergents, etc. is derived directly or indirectly from natural deposits of phosphate. Other sources of phosphorus of minor importance are bones, and basic slag obtained from steel plants. The phosphorus present in rock phosphates and bones is however not used by crop plants effectively because of their exceedingly low solubility in soil solution. The basic purpose of all phosphate manufacturing processes is to produce a material that can dissolve in soil to support plant growth. Justus von Liebig, celebrated as father of agricultural chemistry and chemical fertilizer industry is credited with originating the idea in 1840 that bonemeal could be made more available to plant life by treatment with acid. In May 1842, patents were granted to Lawes and Murray in Great Britain for production of fertilizer by treatment of bones and other phosphatic substances with sulphuric acid. Thus, phosphate industry was born with production of superphosphate. The industry has since diversified.

Classification of Phosphate Fertilizers

Phosphate fertilizers differ in their composition and solubility of phosphorus. Water solubility of phosphorus is the most valued attribute of a phosphate fertilizer but phosphorus present in other forms that are slowly soluble may also be available to plants.

The terms used to describe the phosphorus content in fertilizers are *water soluble, citrate soluble, available and total P* (as P_2O_5). The water soluble fraction is readily soluble in water and citrate soluble fraction is insoluble in water but soluble in neutral ammonium citrate solution. The sum of water-soluble and citrate soluble P represents an estimate of *available P.* The total P represents the total amount of phosphorus present in the fertilizer in any form. The nutrient composition of commercial P fertilizers is given in Table 6.1.

Historically, phosphorus content of fertilizers has been expressed as P_2O_5 (phosphorus pentaoxide), instead of elemental P. Similarly, the potassium concentration in fertilizers is usually expressed as K_2O, instead of K. Although there is no scientific basis for this expression, the fertilizer industry continues to express P and K concentrations in fertilizers as per cent P_2O_5 and per cent K_2O. In fact, neither P_2O_5 nor K_2O is a component of any fertilizer. One kg phosphorus (P) is equivalent to 2.29 kg P_2O_5. Alternately, one kg P_2O_5 is equivalent to 0.43 kg P.

Table 6.1 Nutrient composition of common commercial phosphate fertilizers

Fertilizer	*Analysis (%)*				*% of total P available*
	N	P_2O_5	K_2O	S	
Single superphosphate	–	14–16	–	12–13	97–100
Triple superphosphate	–	46	–	1–2	97–100
Concentrated superphosphate	–	54	–	–	?
Monoammonium phosphate	11	48–52	–	–	100
Diammonium phosphate	18	46	–	–	100
Ammonium phosphate-sulphate	16–20	20	–	13–15	100
Ammonium polyphosphate (liquid)	10–15	34–37	–	–	100
Nitrophosphate	15–23	15–23	0–15	–	80–100
Urea ammonium phosphate	28	28	–	–	100
Rock phosphate	–	20–30	–	–	5–25
Basic slag (Indian)	–	3–8	–	–	70–80
Basic slag (European)	–	12–18	?	–	70–80

Considering the relative availability of fertilizer phosphorus to plants, they are grouped into three categories:

1. **Water-soluble phosphate fertilizers:** This group includes superphosphates and ammonium phosphates, which contain phosphorus in water-soluble form that is readily available to plants. This group of fertilizers is suitable for all crops and in all kind of soils.

2. **Citrate-soluble phosphate fertilizers:** This group includes nitrophosphates which contain major portion of phosphorus in citrate-soluble form and small portion may be present in water-soluble form. Citrate-soluble phosphorus is considered to be slowly available to plants as it dissolves gradually in the soil. This group of fertilizers is most suitable for acid soils.

3. **Insoluble phosphate fertilizers:** This group includes rock phosphates which contain phosphorus in water insoluble form that is partly soluble in citrate solution. The phosphorus present in these fertilizers is not readily available to plants but it reacts with acids in the soil and becomes available gradually. This group of fertilizers is not effective source of phosphorus for crops in neutral and alkaline soils but they can be used in acid soils.

Nearly all phosphate fertilizers are produced directly or indirectly from rock phosphate. For production of phosphate fertilizer, it is necessary to dissolve the insoluble phosphorus mineral present in rock phosphate by treatment with strong acid such as sulphuric acid, phosphoric acid or nitric acid. Ammonia may be used to neutralize the phosphoric acid which is produced from rock phosphate. Thus, there are many possibilities of producing phosphate fertilizers from rock phosphate as shown in Fig. 6.1.

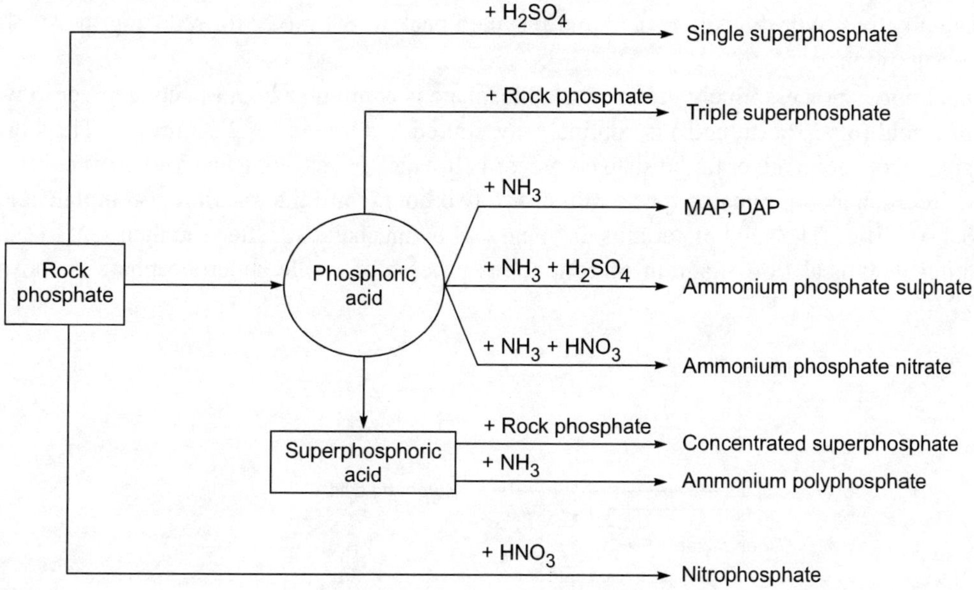

Fig. 6.1 Possibilities of production of different phosphate fertilizers from rock phosphate

SINGLE SUPERPHOSPHATE

Superphosphate is a term used in reference to phosphates which contain phosphorus in a form readily available to plants. Single superphosphate is the first chemical fertilizer which was patented and produced in England in 1843 laying the foundation of chemical fertilizer industry.

Raw Materials

Raw materials required for manufacture of single superphosphate are rock phosphate and sulphuric acid.

Method of Manufacture

There are two processes of manufacture of single superphosphate: batch process and continuous process. In batch process, finely ground rock phosphate of high grade (30-35% P_2O_5) is mixed with sulphuric acid (65-70% strength) in a specially designed mixer for 2-3 minutes. The slurry from the mixer goes to a box type den where the reactions are completed and the material solidifies.

Superphosphate is excavated from the den after 30 minutes to 4 hours and sent to storage for final curing which requires 2-6 weeks depending on the nature and proportion of raw materials and condition of manufacture. The cured product is reclaimed, crushed, screened and bagged or sent to granulator.

$$Ca_{10}(PO_4)_6F_2 + 7H_2SO_4 + 3H_2O = 3Ca(H_2PO_4)_2.H_2O + 7CaSO_4 + 2HF$$

The fluorine compounds generated in the mixer and drier are recovered by scrubbing with water or dilute alkali as hydrofluosilicic acid and recycled back to the mixer for reducing the sulphuric acid consumption.

In continuous process, finely ground rock phosphate is continuously put in cone mixer to which sulphuric acid (65-70% strength) is continuously poured and mixed for 2-3 minutes. The slurry is discharged from cone mixer to the slate conveyor belt which is very long and known as continuous den where the material remains for 30 minutes to two hours until the reaction is complete and the material solidifies. The solid material is disintegrated by mechanical cutter and then sent to storage for curing. A typical flow diagram of continuous process of single superphosphate is shown in Fig. 6.2.

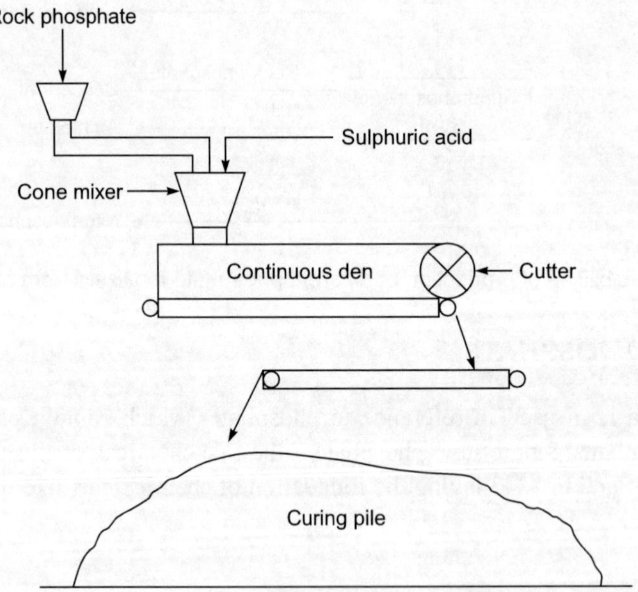

Fig. 6.2 Flow diagram for continuous process of single superphosphate

Product Characteristics

Physical characteristics: Single superphosphate is marketed as a gray material in either powdered or granular form. Powdered material is less expensive but it is dusty and not free-flowing. It has a tendency to form cake on storage. On the contrary, granular product is dust-free and free-flowing, which can be mixed with other fertilizers. The bulk density of granular product is 961 kg/m^3. It is non-hygroscopic with CRH of 93.7% at 30°C.

Chemical characteristics: Single superphosphate is two-fifths monocalcium phosphate and three-fifths gypsum by weight. It contains 14 or 16% P_2O_5, 12% sulphur, and 19-20% calcium. The phosphorus present in single superphosphate is soluble in water. It contains traces of magnesium, iron, aluminium, copper, manganese, zinc and cadmium. The concentration of heavy metals in superphosphate depends directly on the content of these elements in the rock phosphate used as raw material.

TRIPLE SUPERPHOSPHATE

Triple superphosphate is a high analysis fertilizer containing about three times phosphorus than single superphosphate.

Raw Materials

Raw materials required for manufacture of triple superphosphate are rock phosphate and phosphoric acid.

Method of Manufacture

The method of manufacture of triple superphosphate is similar to single superphosphate except that phosphoric acid is used in place of sulphuric acid. Ground rock phosphate is mixed with wet process phosphoric acid (52-54% P_2O_5) in a cone mixer. The slurry from the mixer goes to a den where it solidifies in 10-30 minutes. The product is disintegrated after discharge from the den and piled for curing which requires 3-6 weeks depending on the nature and proportion of raw materials and condition of manufacture. The "run of pile' material is bagged or sent to granulator.

$$Ca_{10}(PO_4)_6F_2 + 14H_3PO_4 + 10H_2O = 10Ca(H_2PO_4)_2.H_2O + 2HF$$

Product Characteristics

Physical characteristics: Triple superphosphate is marketed as a gray material in either powdered or granular form. Powdered material is less expensive but it is dusty and not free-flowing. It has a tendency to form cake on storage. Granular product is dust-free and free-flowing, which can be mixed with other fertilizers. It has the bulk density of 800-881 kg/m^3.

Chemical characteristics: Triple superphosphate contains 46-48% P_2O_5, most of which is monocalcium phosphate monohydrate. It contains little sulphur (1-2%) but substantial amount (2-16%) of calcium. The phosphorus present in triple superphosphate is about 95% soluble in water. It contains traces of magnesium, iron, aluminium, copper, manganese, zinc and cadmium.

CONCENTRATED SUPERPHOSPHATE

Concentrated superphosphate is a high analysis fertilizer containing 54% available P_2O_5. Presently, it is not produced in India.

Raw Materials

Raw materials required for manufacture of concentrated superphosphate are rock phosphate and superphosphoric acid.

Method of Manufacture

Concentrated superphosphate is produced by acidulation of rock phosphate which is mixed with superphosphoric acid diluted to about 73% P_2O_5. Continuous acidulation and partial granulation are accomplished in a rotating drum. Then the product goes to den and curing unit. The "run of the pile" material is screened and bagged.

Product Characteristics

Physical characteristics: Concentrated superphosphate is marketed as a gray material in granular form. The product is dust-free and free-flowing, which can be mixed with other fertilizers. It has bulk density of 1140 kg/m^3.

Chemical characteristics: Concentrated superphosphate is essentially monocalcium phosphate in contrast to monocalcium phosphate monohydrate in triple superphosphate. The loss of a molecule of water, small amounts of fluorine and other volatiles in the production of concentrated superphosphate account for its higher analysis. It contains 54% P_2O_5 and 10-12% calcium. The phosphorus present in concentrated superphosphate is soluble in water.

AMMONIUM PHOSPHATES

Ammonium phosphates are highly concentrated sources of plant nutrients. Two most important ammonium phosphates are monoammonium phosphate and diammonium phosphate.

MONOAMMONIUM PHOSPHATE

Raw Materials

For manufacture of monoammonium phosphate (MAP) wet process phosphoric acid and ammonia are required as raw material.

Method of Manufacture

Anhydrous ammonia is injected into a reactor containing wet process phosphoric acid (45-52% P_2O_5). The ammonia : phosphoric acid mole ratio is maintained at 1 : 1. The reaction is exothermic and releases considerable quantity of steam on completion.

$$NH_3 + H_3PO_4 = NH_4H_2PO_4$$

The concentrated MAP solution flows to spray dryer. The powdered material is collected at the bottom and bagged.

Product Characteristics

Physical characteristics: MAP is a white or light gray crystalline powder with specific gravity of 1.8 and bulk density (1-15 mm size powder) is 849 kg/m^3. It is free-flowing and non-hygroscopic (critical relative humidity is 91.6 at 30°C). It can be handled and stored well without risk of caking.

Chemical characteristics: MAP is a high analysis fertilizer containing 48-52% P_2O_5 and 10-12% N. It is most readily soluble of all phosphatic fertilizers. The pH of 10% solution of MAP is 3.5-4.0.

DIAMMONIUM PHOSPHATE

Raw Materials

For manufacture of diammonium phosphate (DAP) wet process phosphoric acid and ammonia are required as raw material.

Method of Manufacture

In a preneutralizer, anhydrous ammonia is injected into a reactor to partially neutralize wet process phosphoric acid (40-42% P_2O_5). The ammonia : phosphoric acid ratio is maintained at 1 : 4. The resultant slurry containing DAP and MAP is sent to ammoniator-granulator. Ammonia is further added into the bed of granules to increase the mole ratio of ammonia : phosphoric acid to 1 : 2. The reaction is exothermic and heat of reaction is used to remove most of water from granulator. The granules are dried and bagged. Flow diagram for granular diammonium phosphate production is shown in Fig. 6.3.

$$2NH_3 + H_3PO_4 = (NH_4)_2H_2PO_4$$

Product Characteristics

Physical characteristics: DAP is a light gray granular product with bulk density (1-15 mm size powder) of 993 kg/m^3. It is free-flowing and slightly hygroscopic (critical relative humidity is 65-75 at 30°C). It can be handled and stored well with some care.

Chemical characteristics: DAP is a high analysis fertilizer containing 46% P_2O_5 and 18% N. Most of the phosphate is readily soluble in water. The pH of 10% solution of DAP is around 7.8.

AMMONIUM POLYPHOSPHATE

Ammonium polyphosphate (APP) is a high analysis fertilizer that can be produced in liquid or solid form. The most common liquid APP grade is 10-34-0.

Raw Materials

Ammonia and polyphosphoric acid (superphosphoric acid) are required for manufacture of APP.

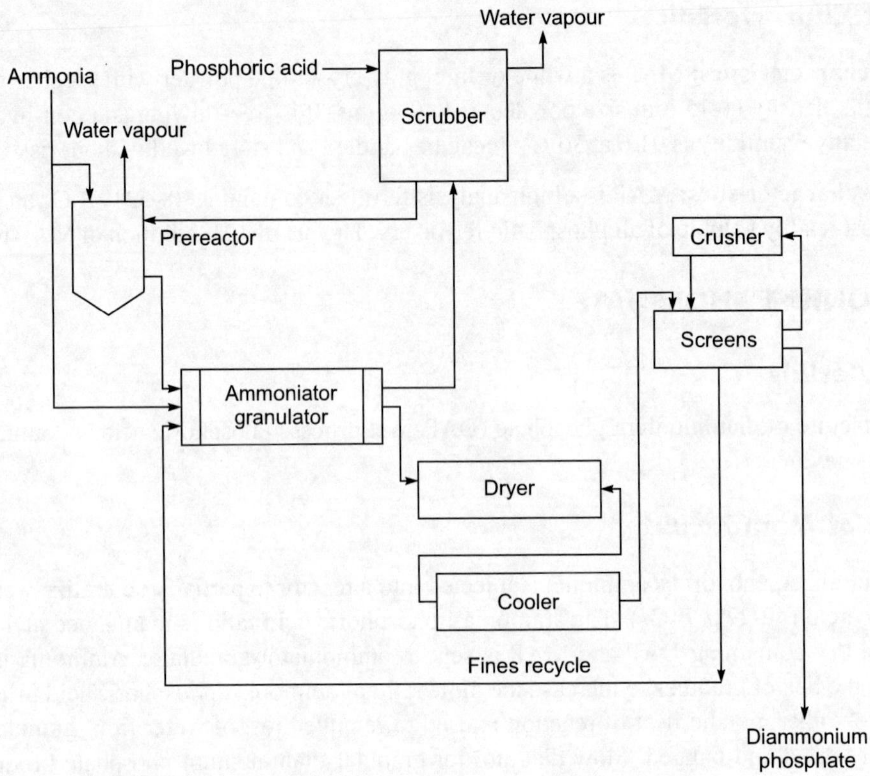

Fig. 6.3 Flow diagram for granular diammonium phosphate production

Method of Manufacture

APP is manufactured by reacting pyrophosphoric acid ($H_4P_2O_7$) with ammonia. Pyrophosphoric acid is produced from dehydration of wet process phosphoric acid. Polyphosphate is a term used to describe two or more orthophosphate ions ($H_2PO_4^-$) combined together, with loss of one molecule of water per two $H_2PO_4^-$ ions.

APP is generally produced in liquid form but it can be produced in solid granular form which is hygroscopic and difficult to handle and use.

Product Characteristics

Liquid APP fertilizer contains 10-15% N and 34-37% P_2O_5 with 75% of the phosphorus present as polyphosphate and remaining 25% as orthophosphate.

One unique property of APP is the sequestering reaction with metal cations. This property maintains a higher concentration of micronutrients in APP than in orthophosphate solutions. For example, APP can maintain 2% zinc in solution compared with only 0.05% zinc in orthophosphate solution.

CALCIUM METAPHOSPHATE

This is the highest grade phosphate fertilizer (65% P_2O_5) which can be produced from rock phosphate by electric furnace method without any acid treatment. Because of high energy requirement and high cost, it is not produced commercially in large quantities.

Raw Materials

Raw material required for manufacture of calcium metaphosphate is rock phosphate alone which is used to produce elemental phosphorus by heating in an electric furnace.

Method of Manufacture

For production of calcium metaphosphate, rock phosphate is heated in an electric furnace to produce elemental phosphorus as discussed in the section on furnace process phosphoric acid. The elemental phosphorus is burnt with air and resultant P_2O_5 fumes are absorbed in finely ground rock phosphate. The reaction between hot phosphorus and rock phosphorus gives calcium metaphosphate in molten form, which solidifies into glassy material on cooling. The glassy material is shattered into small pieces that are ground into fine product.

$$2P_2 + 5O_2 = 2P_2O_5$$

$$Ca_{10}(PO_4)_6F_2 + 7P_2O_5 + H_2O = 10Ca(PO_3)_2 + 2HF$$

Ground limestone is added to prevent lump formation.

Product Characteristics

Physical characteristics: Calcium metaphosphate is a finely divided glassy material with white or faint pale aqua cast. It has a texture similar to fine sand. It flows freely and mixes well with other products. It does not absorb moisture rapidly.

Chemical characteristics: Calcium metaphosphate is the highest grade solid phosphate fertilizer containing 65% P_2O_5, 17% calcium and 3% iron. It also contains 0.2% fluorine. The phosphorus present in calcium metaphosphate is not soluble in water. When applied to humid soils it hydrolyses gradually to orthophosphate ($H_2PO_4^-$) form that is taken up by plants. In acid soils it is almost as effective as superphosphate but in alkaline soils it is less effective than superphosphate.

BASIC SLAG

Basic slag is a by-product of the steel industry where the iron ores contain an appreciable amount of phosphorus. Basic slag is extensively produced and used in Europe especially in Germany, France and England. Small amount of basic slag is also marketed by the steel industry in India as soil conditioner for acid soils.

Method of Manufacture

Before 1960 two popular methods of steel production from the iron ores were Bessemer process and open hearth process. Today, these processes have been largely replaced by basic oxygen process. Phosphorus is removed from iron ore because its presence weakens the cast iron that is produced from pig iron. The steel production process involves heating iron ore and lime in a furnace. When the mixture becomes red hot, the lime melts and combines with the acid impurities including phosphorus. The resulting compounds containing calcium, phosphorus, silicon, magnesium and manganese are lighter than iron and they rise to surface of the molten mass and are skimmed off as slag. After cooling the slag solidifies to a hard, black massive cinder, which is crushed and ground to a fine powder.

Product Characteristics

Physical characteristics: Basic slag is heavy, dark brown powder which is ground to pass 70% through a 100-mesh sieve. It is dusty material which may be mixed with pulverized fertilizer to improve its handling characteristics. It is likely to solidify if stored in a damp place.

Chemical characteristics: European basic slag contains higher amount of phosphorus, usually 12-18% P_2O_5 while Indian basic slag contains low amount, usually 3-8% P_2O_5. The phosphate content of basic slag depends on the phosphorus content of the iron ore used for steel production. Indian iron ores have low phosphorus content.

The phosphorus in basic slag is insoluble in water but is mostly soluble in 2% citric acid. About 70-80% of the total P present in the basic slag is soluble in citric acid and becomes slowly available in acid soils. It is not recommended for neutral and alkaline soils.

Basic slag is alkaline in nature and may act as good liming material for acid soils. It has neutralizing value of 70% lime equivalent ($CaCO_3$). It contains magnesium, iron and manganese with traces of boron.

ROCK PHOSPHATE

Rock phosphate is the raw material for manufacture of almost all phosphate fertilizers. The characteristics and typical composition of selected rock phosphates have been discussed in Chapter 2.

Since the phosphorus present in rock phosphate is insoluble in water, its use as fertilizer for direct application is not common simply because it is not effective P source for crops. Recently, there has been some interest in use of some reactive rock phosphates for direct application to acid soils as phosphate source in developing countries. The advantages are very low unit cost of phosphate and possibility of utilizing low grade rock phosphates which are not suitable for use as raw material for production of phosphoric acid and superphosphates. The interest stems from the results of extensive research which has revealed that rock phosphate can be used directly as P fertilizers under certain conditions, however, in most situations P fertilizers processed from rock phosphate are more cost-effective. The effectiveness of rock phosphate depends entirely on its dissolution in the soil, which is mainly influenced by its chemical reactivity and soil properties. Soil acidity, exchangeable calcium and aluminium and organic matter content influence the dissolution of rock phosphate. Strong acidity, low calcium and aluminium and high organic matter content favour the dissolution of rock phosphate in the soil.

There is wide variability in the agronomic effectiveness of rock phosphates. While some rock phosphate sources are not suitable for direct application, some others can be used with advantage in acid soils. The selection of the effective rock phosphate can be made on the basis of the reactivity measured as solubility in neutral ammonium citrate or 2% citric acid. The citrate solubility varies usually from 5 to 25% of the total P. The results of reactivity measurement for some selected rock phosphates are presented in Table 6.2. Significant differences among rock phosphate sources was observed in regard to the reactivity with Indian sample testing the lowest value. Rock phosphates with solubility in neutral ammonium citrate and 2% citric acid of more than 18 and 40% of the total P, respectively, are considered highly reactive and almost as effective as water soluble P fertilizers under favourable conditions.

Table 6.2 Solubility of selected rock phosphates (ground 100% < 100 mesh)

Source of rock phosphate	*Solubility of* P_2O_5 *(% of the rock)*	
	Neutral ammonium citrate	*2% citric acid*
Gafsa (Tunisia) unground	6.3	10.6
North Carolina (USA)	7.1	15.9
Hahotoe (Togo)	4.1	7.6
Kaiyang (China)	3.4	7.2
Araxa (Brazil)	2.8	5.0
Jhamarkotra (India)	0.6	1.5

Finely ground rock phosphates are effective as phosphorus source for crops only in acid soils with pH less than 6.0. It should be applied three to five times the amount of conventional water soluble P fertilizer and thoroughly mixed into the soil. Environmental conditions such as warm climate, moist soils and long growing seasons will increase the effectiveness of rock phosphate. It is therefore extensively used for plantation crops such as rubber, oil palm and coconut on very acid soils (pH less than 5).

Factors limiting the use of rock phosphate include uncertain agronomic value, inconvenience of handling and applying the fine dusty material and relatively low P content compared with DAP and TSP.

Selected References

FAI (2001). *Handbook on Fertilizer Technology*. 6th ed., Fertiliser Association of India, New Delhi. pp. 273.

Havlin, J.L., Beaton, J.D., Tisdale, S.L. and Nelson, W.L. (2005). *Soil Fertility and Fertilizers*, 7th ed. Prentice-Hall of India, New Delhi.

Mordtvet, J.J., Murphy, L.S. and Follett, R.F.H. (1999). *Fertilizer Technology and Application.* Meister Publishing Company, Euclid Ave., Willoughby, OH.

UNIDO/IFDC, Ed. (1998). *Fertilizer Manual*, 3rd ed., Kluwer Academic Publishers, Dordrecht, the Netherlands.

CHAPTER

7

Potash Fertilizers

The word 'potash' which is a trade name commonly applied to potassium-containing fertilizers is derived from pot ashes. During early days, wood and other plant residues were burnt in pots to obtain salts which were required for various purposes including soap manufacture and glass-making. The salts were recovered from ashes by extraction with water followed by evaporation of the solution. The residue consisted of a mixture of potassium carbonate and other salts. Until middle of nineteenth century, wood ashes remained the chief source of potassium. First major source of potassium in natural deposits was discovered during exploration of salt deposits in Strassfurt, Germany in 1851. The first factory came up in 1861 for processing crude potassium chloride. Since then numerous mineral deposits and brine sources of potassium have been discovered and processing plants are scattered over much of the world. In 2006, world production of potassium was over 30 million tonnes.

POTASH FERTILIZERS

Common potash fertilizers are listed in Table 7.1. All potash fertilizers are water-soluble. They consist essentially of potassium in combination with chloride, sulphate or nitrate. Some double salts such as sulphate of potash-magnesia also exist.

Like phosphorus, the potassium content of fertilizers is guaranteed in terms of potassium oxide (K_2O) equivalent, though potassium oxide is not present in any commercial fertilizer. For converting % K into % K_2O, the following expressions are used.

$$\% K_2O = \% K \times 1.20$$

$$\% K = \frac{\% K_2O}{1.20}$$

Table 7.1 Common commercial potash fertilizers

Fertilizer	*Plant nutrient content (%)*				
	N	P_2O_5	K_2O	S	Mg
Muriate of potash (Potassium chloride)	–	–	60–62	–	–
Sulphate of potash (Potassium sulphate)	–	–	50–52	17	–
Sulphate of potash-magnesia (Potassium-magnesium sulphate)	–	–	22	22	11
Potassium nitrate	13	–	44	–	–

MURIATE OF POTASH

About 95% of world production of potassium fertilizers is in the form of Muriate of Potash, also known as potassium chloride. The word "muriate" is derived from the term "muriatic acid" which is old name of hydrochloric acid. Muriate of potash is a potassium salt of hydrochloric acid.

Raw Materials

Muriate of potash is commercially produced from either sylvinite ore or brines. A description of the natural deposits of commercially important potash ores in the world is given in Chapter 2.

Methods of Manufacture

Three methods are followed for recovery of potassium chloride from sylvinite ore, viz.

- fractional crystallization,
- froth flotation, and
- heavy media separation.

Fractional Crystallization Process

This is an old and popular process in which sodium chloride (NaCl) is removed by solution and recrystallization. The process is based on the difference in temperature-solubility relationship of potassium chloride and sodium chloride. The solubility of potassium chloride decreases rapidly with fall in temperature whereas the solubility of sodium chloride changes only slightly over a wide temperature range and below 95°C, it is in retrograde (Fig. 7.1). Brines saturated with both sodium chloride and potassium chloride below 95°C, therefore, are found most suitable for vacuum crystallization.

In this process, potassium ore is mixed with hot (about 100°C) brine, saturated with respect to sodium chloride, in such a ratio that all of potassium chloride is dissolved from the ore resulting in a solution saturated with potassium chloride. Sodium chloride is not soluble in this condition, so it remains in the solid phase and is discarded. The hot solution is then cooled to ambient temperature causing potassium chloride to crystallize while sodium chloride remains in solution. Vacuum

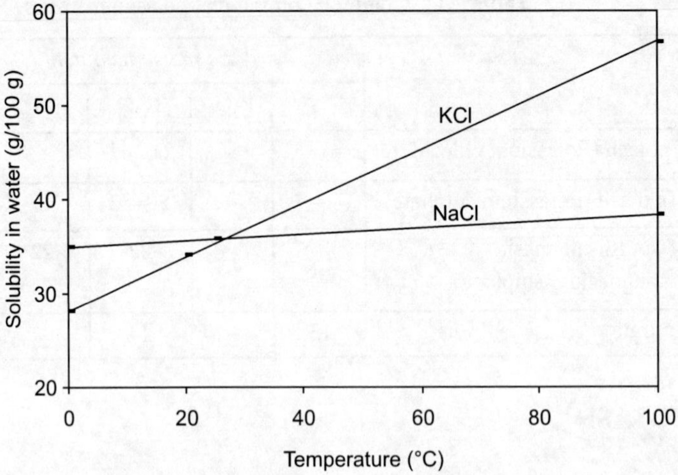

Fig. 7.1 Effect of temperature on the solubility of potassium chloride and sodium chloride in water

crystallizers are generally used to produce large size crystals. After separation of potassium chloride crystals, mother liquor is heated and recycled back to ore extraction unit. A simplified flow diagram showing various steps involved in the production of muriate of potash by fractional crystallization process is shown in Fig. 7.2.

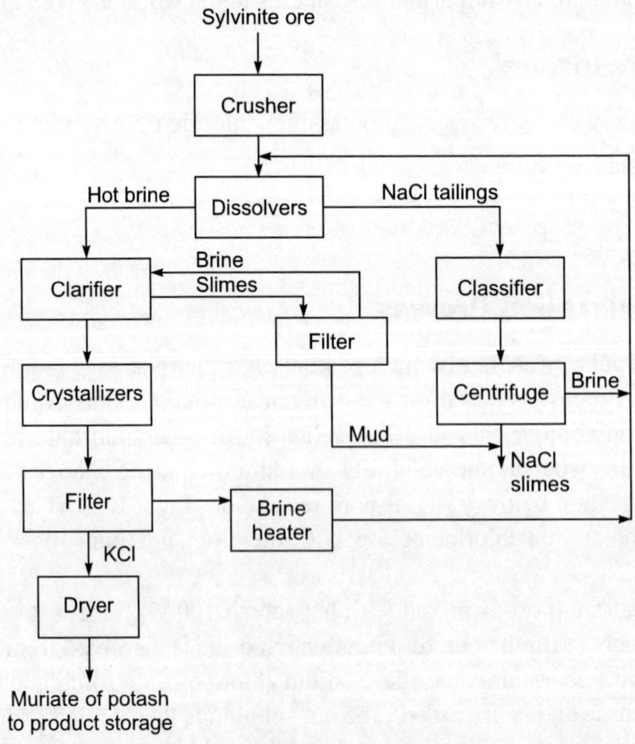

Fig. 7.2 Simplified flow diagram for crystallization process of muriate of potash

Froth Flotation Process

This is a mechanical separation process by which solid particles of one or more types of minerals which are suspended in slurry are selectively caused to rise to the surface of the container by the action of air bubbles. The floating particles are removed by skimming off the froth. This process entails six steps:

(i) Grinding of the ore to separate the crystals of sylvite (KCl) from halite (NaCl) and clay slime.
(ii) Suspending the ground ore in brine saturated with KCl and NaCl.
(iii) Desliming to remove clay from the slurry.
(iv) Conditioning of the deslimed slurry with proper auxiliary and flotation agents to film sylvite crystals selectively. Auxiliary agent or blinding agents are used to prevent the remaining clay from combining with expensive flotation agent. Guar starch or gums are used to act as binding agent. Primary aliphatic amine acetate salts or sulphonates derived from beef tallow are widely used as flotation agent. A typical flotation agent is made of approximately 25% mono-n-octadecyl, 25% mono-n-hexadecyl and 50% mono-n-octadecenyl neutralized with glacial acetic acid.
(v) In the flotation cell air is drawn by agitation. The air bubbles attach with the filmed sylvite crystals and float them to the surface. The froth containing sylvite crystals is mechanically skimmed off by paddles.
(vi) Recovery of potassium chloride from overflows or flotation tailings by leaching, thickening and recystallization.

A simplified flow diagram of the froth flotation process is shown in Fig. 7.3. Froth flotation process is more popular than traditional crystallization process. This method is also used for separation of potassium chloride from sodium chloride in the salt mixture obtained from brine sources.

Heavy Media Separation

This is a new process developed for separation of potassium chloride from sodium chloride in coarse ore (+10 mesh screen). In this process, saturated brine is mixed with enough magnetite (specific gravity 5.7) to adjust the specific gravity of the mixture to a value between 1.99 (specific gravity of KCl) and 2.17 (specific gravity of NaCl). The coarse fraction is blended with brine containing magnetite and pumped through hydrocyclones which act as centrifugal separators. Potassium goes in the overflow whereas sodium chloride remains in the underflow. Magnetite is recovered from both the salts. Potassium concentrate is debrined, dried, crushed and screened.

Extraction from Brine

There are few plants which extract potassium salts from brines obtained from a lake or sea or underground well. Commercially important salt lakes for potassium extraction are Searle's Lake, California; the Great Salt Lake, Idaho; and the Dead Sea. The plant at Searle's Lake, California is in operation for over 100 years. Here the brine from 25 to 40 metres above the bottom of the

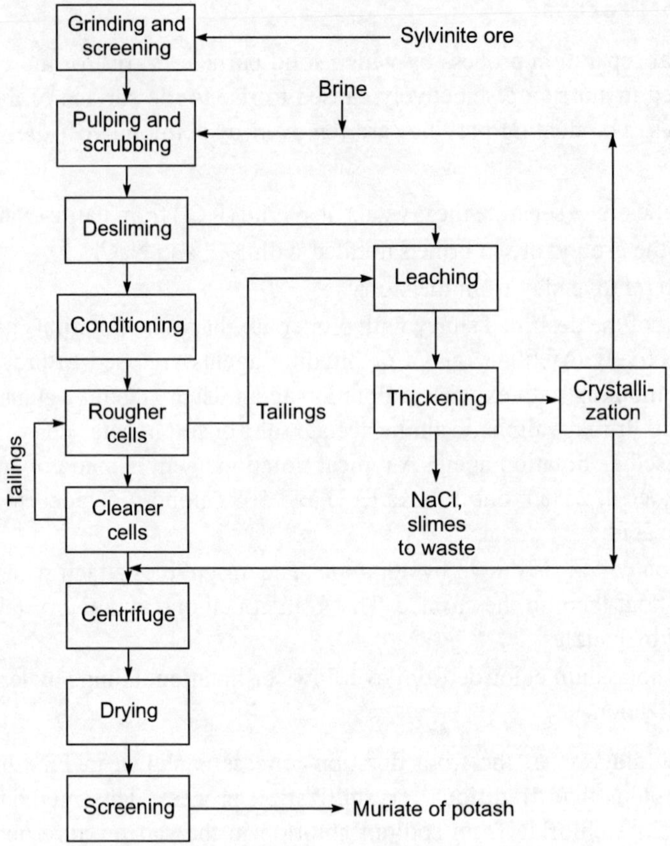

Fig. 7.3 Simplified flow diagram for flotation process of potash recovery

lake is pumped several kilometres to the processing plant. Recovery of potassium chloride is accomplished by a process of evaporation, followed by fractional crystallization which is known as Trona process. Raw brine is first circulated through coolers to serve as cooling medium and at the same time absorbing heat for its own processing. Hot brine is mixed with recycled process liquor and concentrated in big triple effect evaporators under vacuum. As evaporation proceeds, burkeite ($2Na_2SO_4.Na_2CO_3$) and sodium chloride precipitate are removed through salt traps. The hot concentrated liquor after separation of sodium salts is clarified and sent to potash plant where potassium chloride is separated from borax in three-stage vacuum coolers.

In Israel and Jordan potassium is extracted from Dead Sea brine which contains about 30% salts (4.4% potassium chloride) using nature's intense desert solar energy and heat. The extraction process starts by the evaporation of the Dead Sea water in the evaporation ponds. Fertilizer grade potassium chloride is manufactured from the salts collected from evaporation ponds.

From the Dead Sea brine, Israel produces 1.8 million tonnes potash, 206,000 tonnes elemental bromine, 44,900 tonnes caustic soda, 25,000 tonnes magnesium metal, and sodium chloride. On the Jordanian side of the Dead Sea, the *Arab Potash Co.* produces 2.0 million tonnes of potash annually, as well as sodium chloride and bromine. The process involves use of extensive salt evaporation pans that have essentially diked the entire southern end of the Dead Sea for the purpose of producing carnallite (potassium magnesium chloride) which is then processed further to produce potassium chloride.

Secondary Processing

In some developed countries where bulk blending industry is prominent, potash products in coarse and granular form are in demand. This stems from the problem of segregation of fertilizer blends, which can be minimized when coarse muriate of potash compatible with granular nitrogen and phosphate materials is used. To achieve this, two methods that are followed for converting fine muriate of potash into coarse product are fusion and compaction.

In ***fusion method***, finely divided muriate of potash is heated to 755°C to form a melt which is solidified by cooling. The flakes thus formed are crushed and screened to give the desired particle size.

In ***compaction method*** which is more widely used, muriate of potash is fed between two pressure rollers (weight 300 tonnes) at temperature ranging from 93 to 121°C in order to compress it into sheets (0.16 to 0.32 cm) or briquettes. Briquettes are crushed and screened to desired size.

Product Characteristics

Physical characteristics: Muriate of potash is white or greyish red crystalline solid. Sometimes, it may look like pepper and salt mixture. Product made by fractional crystallization method is white in colour whereas that produced by flotation process is usually pink to red in colour due to presence of small quantity of coloured impurities.

Fertilizer grade potassium chloride is usually white. It is slightly hygroscopic (critical relative humidity at 30°C is 84) and free-flowing. In developed countries granular product is available which is dust-free and free-flowing that stores well. It can be mixed with nitrogen and phosphorus fertilizers to produce multinutrient fertilizers.

Chemical characteristics: Fertilizer grade potassium chloride contains 60% K_2O although some sodium chloride is present as impurity. Fertilizer grade product contains traces of bromine, sulphur, magnesium, calcium, iron and boron.

It is a water-soluble product. At 20°C, saturation of water with potassium chloride is reached at 320 g/L.

POTASSIUM SULPHATE

Potassium sulphate is used in specialty fertilizers as its unit cost is greater than muriate of potash. For some crops like grapevine, tobacco, etc. that are sensitive to excess chloride, potassium sulphate is preferred over potassium chloride.

Raw Materials

Potassium sulphate is produced from a number of raw materials depending on the process of manufacture. However, potassium chloride is commonly the source of potash in these processes. The raw materials required in important processes are:

- Potassium chloride and langbeinite in Langbeinite process
- Potassium chloride and burkeite in Trona process
- Potassium chloride and sulphuric acid in Mannheim process
- Potassium chloride and sulphur in Hargreaves process

Method of Manufacture

Potassium sulphate is produced by several processes. Four commercially important processes are briefly discussed here:

Langbeinite process: In this process langbeinite, a double sulphate of potassium and magnesium ($K_2SO_4.2MgSO_4$) is dissolved in water and subsequently concentrated solution of potassium chloride is added. The following reaction yields a precipitate of potassium sulphate which is separated by centrifuging.

$$K_2SO_4.2MgSO_4 + 4KCl = 2K_2SO_4 + 2MgCl_2$$

The wet cake is dried, screened and sent to storage.

Trona process: In this process, potassium chloride is first reacted with burkeite ($Na_2CO_3.2Na_2SO_4$) to produce a high potassium glaserite cake ($Na_2SO_4.3K_2SO_4$). The glaserite is separated and then treated with high purity potassium chloride brine. This yields solid potassium sulphate which is filtered from the NaCl brine.

$$\underset{\text{Burkeite}}{Na_2CO_3.2Na_2SO_4} + 4KCl = 2K_2SO_4 + 4\,NaCl + Na_2CO_3$$

Potassium sulphate is washed, dried and sent to bulk storage. A simplified flow diagram of Trona process for potassium sulphate production is shown in Fig. 7.4.

Mannheim process: This process involves reaction between potassium chloride and sulphuric acid. The reaction proceeds in two stages. In the first stage, which is exothermic, potassium acid sulphate is formed.

$$KCl + H_2SO_4 = KHSO_4 + HCl$$

In the second stage, which is endothermic, the acid sulphate is converted into potassium sulphate by further reaction with potassium chloride.

$$KHSO_4 + KCl = K_2SO_4 + HCl$$

In the industry, both stages of the reaction are conducted in a Mannheim-type salt cake furnace. Potassium chloride and sulphuric acid are fed in separately and continuously from the top and near the centre. As the mixture moves outward through hotter zones (600–700°C) the reaction is completed.

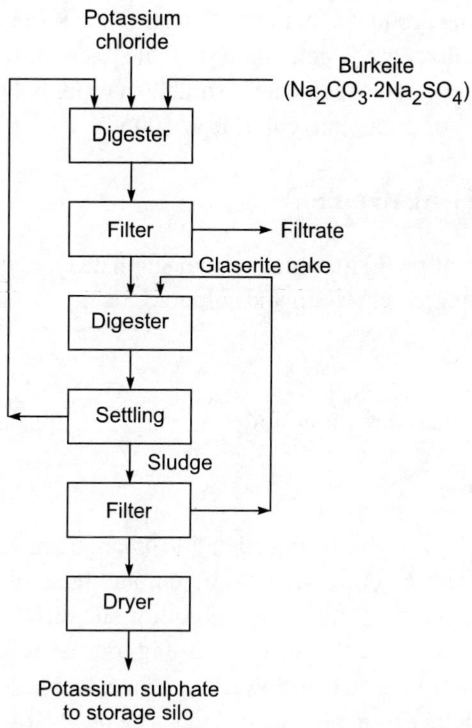

Fig. 7.4 Simplified flow diagram of Trona process for potassium sulphate production

The hot potassium sulphate coming out of furnace is cooled and then bagged. The hydrogen chloride gas passes out of the furnace at the top and is absorbed in water to form hydrochloric acid.

The potassium sulphate produced by this process contains little free acid.

Hargreaves process: In the process gas containing sulphur dioxide, excess air and water vapour is passed through beds of potassium chloride briquettes. The reaction of potassium chloride with sulphur produces potassium sulphate.

$$2KCl + SO_2 + H_2O = K_2SO_4 + 2HCl$$

The reaction is endothermic and temperature control is critical. The converted potassium sulphate briquettes are cooled, removed from the chamber, crushed and screened.

Product Characteristics

Physical characteristics: The physical and handling characteristics of potassium sulphate are better than potassium chloride because it is less hygroscopic (CRH at 30°C is 96.3). It is colourless or white product which is available in standard or granular grades. In the standard grade it mixes well with other fine ingredients of mixed fertilizers. In granular grade it is adapted to bulk blending and granular mixtures.

Chemical characteristics: The product contains 50% K_2O (42% K) and 16% sulphur and not more than 2.5% chlorine. In the fertilizer grade potassium sulphate traces of sodium, calcium, magnesium and bromine are also present. Potassium sulphate is soluble in water but less than potassium chloride (24.1% as compared to 56.7% of potassium chloride at 100°C).

SULPHATE OF POTASH-MAGNESIA

This fertilizer is a double salt of potassium-magnesium sulphate ($K_2SO_4.2MgSO_4$) which provides three macronutrients: potassium, magnesium and sulphur.

Raw Materials

Langbeinite is the only raw material required for production of sulphate of potash-magnesia.

Method of Manufacture

Sulphate of potash-magnesia is produced by beneficiation of langbeinite. The ore contains potassium magnesium sulphate mixed with chloride salts of sodium and potassium. The latter salts dissolve at a much faster rate than the sulphate salts. This solubility rate difference is used to advantage in beneficiating the ore. A continuous countercurrent washing process is employed.

The ore is first crushed and screened followed by controlled washing to dissolve away the chloride salts leaving langbeinite essentially unchanged. The residual solid phase is separated from the wash water by centrifuging and dried.

Product Characteristics

Physical characteristics: Sulphate of potash-magnesia is a free-flowing colourless product with excellent physical and handling qualities. It is offered as granular or powdered material which can be used in blending, mixing or other solid fertilizers manufacture.

Chemical characteristics: Fertilizer grade sulphate of potash-magnesia contains 96–98% langbeinite. It contains three major nutrients: 21–22% K_2O, 11–18% magnesium and 22–23% sulphur. The standard material should not contain more than 2.5% chlorine.

POTASSIUM NITRATE

Potassium nitrate is also known as niter or *saltpeter*. Though it is one of the oldest fertilizer materials used in agriculture, it is now produced commercially in limited quantities because of high cost of manufacture as compared with other potash fertilizers.

Raw Materials

Historically, potassium nitrate was recovered from natural 'caliche' deposits in Chile. Most of the caliche deposits have already been exploited and only a small quantity is produced now. Today, synthetic potassium nitrate is produced from potassium chloride and nitric acid. In China potassium nitrate is produced from potassium chloride and ammonium nitrate.

Method of Manufacture

Potassium nitrate is now produced by a chemical reaction between nitric acid and muriate of potash. The process consists of combining potassium chloride and nitric acid in a packed tower.

$$4KCl + 4HNO_3 = 4KNO_3 + 2Cl_2 + 2H_2O$$

Molten potassium nitrate is removed from bottom of the tower and the gaseous by-products (chlorine gas, water vapour) are removed from the top.

In China, potassium nitrate is produced by reacting potassium chloride with ammonium nitrate in the secondary mother liquor from which potassium nitrate (KNO_3) and ammonium chloride (NH_4Cl) have been separated. The reaction liquid goes to the intermediate tank where it is cooled and then it is sent to a vacuum cooling crystallizer, where most of KNO_3 are crystallized out. A centrifuge separates out the product KNO_3. The primary mother liquor so obtained is then sent into cold-precipitating crystallizer to allow NH_4Cl to crystallize. The slurry then goes through a centrifuge to separate out the by-product NH_4Cl.

$$KCl + NH_4NO_3 = KNO_3 + NH_4Cl$$

Physical characteristics: Fertilizer grade potassium nitrate is colourless crystalline material. It is non-hygroscopic (CRH at 30°C is 90.5). It can be mixed well with other fertilizers.

Chemical characteristics: Fertilizer grade potassium nitrate contains 13% nitrogen and 44% K_2O (36% K). Potassium nitrate is soluble in water to the extent of 13% at 0°C to 246% at 100°C. In cold water it is less soluble than muriate of potash but more soluble than potassium sulphate. At room temperature, potassium nitrate and muriate of potash exhibit about the same solubility, which is much higher than potassium sulphate. In water above room temperature, the solubility of potassium nitrate is not only greater than that of muriate of potash and potassium sulphate but it also increases rapidly as the temperature rises while those of muriate of potash and potassium sulphate remain relatively unchanged.

Potassium nitrate is used to pickle meat and in the manufacture of pyrotechnics, explosives, matches, rocket propellants.

Selected References

Follett, R.F.H., Murphy, L.S. and Donahue, R.L. (1981). *Fertilizers and Soil Amendments*. Prentice-Hall Inc., Englewood Cliffs, N.J.

Havlin, J.L., Beaton, J.D., Tisdale, S.L. and Nelson, W.L. (2005). *Soil Fertility and Fertilizers*, 7th ed. Prentice-Hall of India, New Delhi.

Mordtvet, J.J., Murphy, L.S. and Follett, R.F.H. (1999). *Fertilizer Technology and Application*. Meister Publishing Company, Euclid Ave., Willoughby, OH.

UNIDO/IFDC, Ed. (1998). *Fertilizer Manual*, 3rd ed., Kluwer Academic Publishers, Dordrecht, the Netherlands.

CHAPTER

8

Complex and Mixed Fertilizers

There are two types of fertilizers: straight fertilizers which contain only one of the major nutrients, and multinutrient fertilizers which contain two or more major nutrients, usually NP or NPK. Complex fertilizers are products obtained by chemical reaction of two or more fertilizer materials and contain two or more major plant nutrients. Their composition is complex consisting of two or more chemical compounds. Mixed fertilizers are also multinutrient fertilizers but their production does not involve any significant reaction and they are essentially physically mixed products. Production and use of complex and mixed fertilizers have several advantages such as:

1. They are high-analysis fertilizers containing more than 30% plant nutrients.
2. They are granular and free-flowing with uniform grain size.
3. They supply N and P or N, P and K in available form in one operation, thus assure balanced nutrition of crops.
4. Their cost per kg nutrient (unit cost) is lower than straight fertilizers.
5. Due to their high analysis, transport and distribution cost per unit nutrient is less than in straight fertilizers.
6. They are non-caking and non-hygroscopic, thus safe for handling and storage.

Recently, use of complex and mixed fertilizers has been growing rapidly. The value of complex or mixed fertilizers depends on the following factors:

1. The content of individual nutrients: N, P and K.
2. The ratio of N to P_2O_5 or N to K_2O.
3. The form in which individual nutrients are present.
4. The residual effects on soil, i.e. acidity or basicity of the material.
5. The content of secondary and micronutrients.

Nutrient content is the most important factor determining the value of a complex or mixed fertilizer because the production cost as well as transportation and handling costs per unit of nutrient decrease with increase in the nutrient content. High analysis products are therefore preferred over low analysis products.

The ratio of nutrients (N : P and N : K) determines the suitability of the multinutrient fertilizer for the crops in a given area. A higher proportion of P and K than N is preferred because P and K are applied only once at the time of planting, whereas N is applied in two-three split doses (basal and top dress in the standing crop).

Generally, the nutrients in the standard fertilizers are present in such chemical forms that are readily available to plants. However, use of inexpensive sources of nutrients like rock phosphate and basic slag in production of multinutrient mixed fertilizers make the product low priced but less effective.

The residual effects of complex fertilizers are of significance only when the same product is used for a long time, especially in acid soils.

The presence of secondary and micronutrients in the complex and mixed fertilizers add value and make them attractive to the farmers.

COMPLEX FERTILIZERS

Ammonium Phosphate Sulphate

Ammonium phosphate sulphate is composed mainly of ammonium phosphate and ammonium sulphate with a typical grade of 16-20-0. Other grades of this product are also produced.

Raw Materials

For production of ammonium phosphate sulphate, the raw materials required are ammonia, phosphoric acid and sulphuric acid. Urea is also required for production of 20-20-0 grade product.

Method of Manufacture

For production of 16-20-0 grade ammonium phosphate sulphate, a mixture of phosphoric acid (25-30% P_2O_5) and sulphuric acid is directly neutralized with ammonia. The resulting slurry is pumped to the blunger for granulation. This produces a mixture containing 42% monoammonium phosphate and 58% ammonium sulphate. In another process, ammonium sulphate solution is added to phosphoric acid and then the mixture is ammoniated. Urea can be added in the blunger for the production of 23-23-0 grade. Fig. 8.1 depicts a simplified flow diagram for ammonium phosphate sulphate production.

Product Characteristics

Physical characteristics: Ammonium phosphate sulphate is granular product of grey colour which is free-flowing. It can be stored well without any risk of caking.

Chemical characteristics: Ammonium phosphate sulphate is a high analysis fertilizer that is produced in two grades, viz. 16-20-0 and 23-23-0. It also contains 13-15% sulphur in sulphate form. All the nutrients are readily soluble in water.

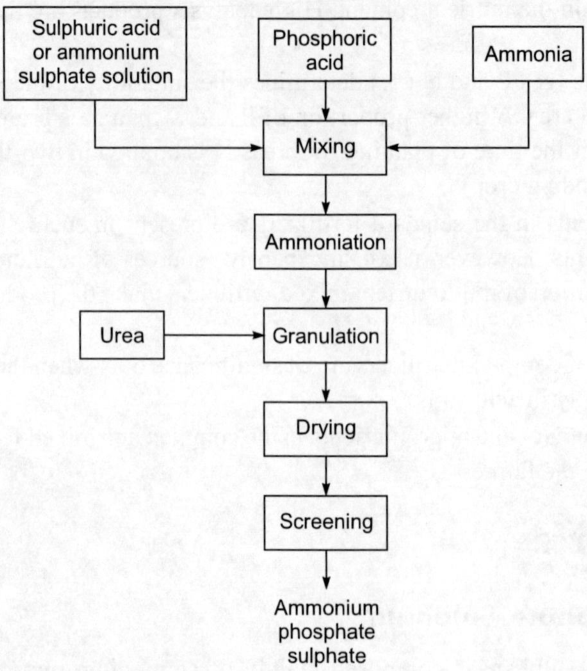

Fig. 8.1 Flow diagram for ammonium phosphate sulphate production

UREA-AMMONIUM PHOSPHATE

Urea-ammonium phosphate is high analysis multinutrient fertilizer with acceptable characteristics.

Raw Materials

Phosphoric acid, ammonia and urea are required as raw materials for production of urea-ammonium phosphate.

Method of Manufacture

Ammonia is introduced in preneutralizer containing wet process phosphoric acid. The resulting ammonium phosphate slurry is pumped to the granulator. In the granulator more ammonia and urea are fed to increase N content. Some filler like sand or dolomite may be added as desired depending on the grade. The finished granules are dried, screened, cooled and coated with anti-caking agent.

Product Characteristics

Physical characteristics: Urea-ammonium phosphate (UAP) is granular product of gray or cream colour which is free-flowing and hygroscopic (critical relative humidity is 57 at 30°C). It needs careful handling and storage.

Chemical characteristics: UAP is a high analysis fertilizer that is produced in three grades, viz. 28-28-0, 24-24-0 and 20-20-0. About one-third of the total N is in ammoniacal form and remaining in the form of urea. Most of phosphate is readily soluble in water.

NITROPHOSPHATES

The term nitrophosphate covers a number of fertilizers containing nitrogen and phosphorus and sometimes potassium too, which are produced by the reaction of nitric acid with rock phosphate. Ammonium nitrate is one of the principal components in the product. The nitrophosphate process is sulphur independent and was developed in Europe due to shortage of sulphur. It is more energy efficient, has raw material flexibility and is without by-product disposal problem.

Raw Materials

The important raw materials required to produce nitrophosphate are nitric acid, phosphoric acid, rock phosphate and ammonia. Depending on the process adopted, other materials like sulphuric acid, DAP and ammonium sulphate are also added. For production of NPK grades, potassium chloride is also required.

Methods of Manufacture

The basic principle of nitrophosphate manufacture is the acidulation of ground rock phosphate with nitric acid (53-60% concentration) in a series of reactors. The reaction mixture contains calcium nitrate and phosphoric acid. The reaction may be expressed as follows:

$$\underset{\text{Fluorapatite}}{Ca_{10}(PO_4)_6F_2} + 14HNO_3 + 3H_2O \rightarrow \underset{\text{Monocalcium phosphate}}{3Ca(H_2PO_4)_2.H_2O} + \underset{\text{Calcium nitrate}}{7Ca(NO_3)_2} + 2HF$$

$$\underset{\text{Monocalcium phosphate}}{Ca(H_2PO_4)_2} + \underset{\text{Calcium nitrate}}{Ca(NO_3)_2} + 2\ NH3 \rightarrow \underset{\text{Dicalcium phosphate}}{2CaHPO_4} + \underset{\text{Ammonium nitrate}}{NH_4NO_3}$$

$$Ca(NO_3)_2 + 2\ NH_3 + CO_2 + H_2O \rightarrow 2\ NH_4NO_3 + CaCO_3$$

The calcium nitrate that is formed is very hygroscopic, hence it must be removed or altered to improve the product quality. The nitrophosphate processes differ in the manner in which calcium nitrate is dealt with. Commercial processes for production of nitrophosphate may be grouped into four categories:

(i) Processes in which calcium nitrate is separated by cooling, e.g. ODDA, Norsk Hydro, BASF.

(ii) Sulphonitric process in which calcium nitrate is replaced with calcium sulphate by use of sulphuric acid or sulphate salts.

(iii) Phosphonitric process in which calcium nitrate is replaced with monocalcium phosphate by use of phosphoric acid or phosphate salts.

(iv) Carbonitric process in which calcium nitrate is replaced with calcium carbonate by use of carbon dioxide. This process is simple but not popular as it yields a product that contains nearly all P in water insoluble form.

The mixture can be converted into a solid granulated fertilizer by any of four methods.

Recent advances in nitrophosphate production technology have made it possible to produce nitrophosphate based NP and NPK fertilizers with provision for conversion of calcium nitrate into calcium ammonium nitrate (CAN) fertilizer. Ammonium nitrophosphate (20-20-0) is such a product with CAN as co-product. The flow diagram for production of ammonium nitrophosphate is shown in Fig. 8.2.

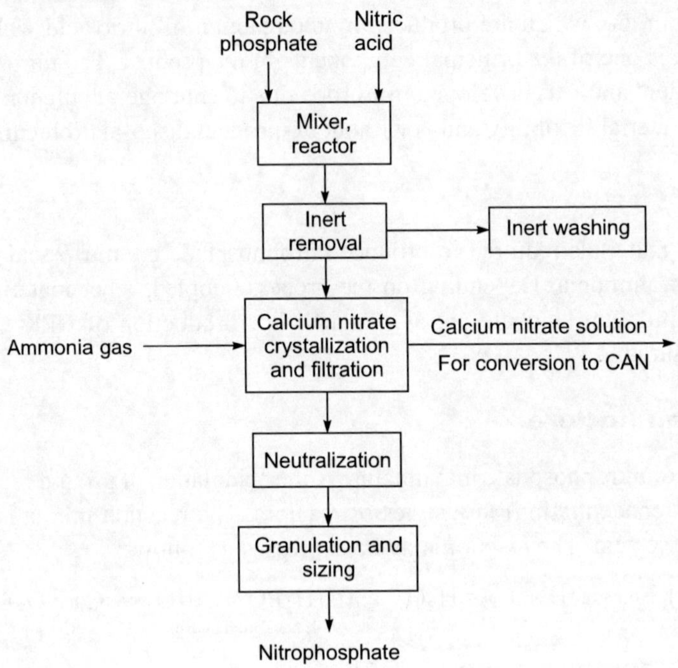

Fig. 8.2 Flow diagram for nitrophosphate (ANP) production

Product Characteristics

Physical characteristics: Nitrophosphate based fertilizers are granular or in prill form. They are free-flowing. The bulk density is around 1100 kg/m^3. It is however hygroscopic with CRH of 57 and 50 at 30°C for 20-20-0 and 15-15-15 grades, respectively. The product can however be stored for prolonged period under normal warehousing conditions without risk of caking.

Chemical characteristics: Nitrophosphate fertilizers are high analysis fertilizers with grades such as 20-20-0, 23-23-0, 15-15-15. The nitrogen content is in partly ammoniacal and partly nitrate form. About 98% of the phosphorus in nitrophosphate is soluble in neutral ammonium citrate, and 30-50% is soluble in water.

NPK COMPLEX FERTILIZERS

NPK complex fertilizers are solid fertilizers produced as high analysis granular products. Several grades of NPK complex fertilizers are produced in India such as 15-15-15, 17-17-17, 10-26-26,

12-32-16, 22-22-11, 14-35-14, 14-28-14, 19-19-19 and 20-10-10. These fertilizers are convenient to use because they contain all the three major plant nutrients (NPK) in the desired proportions.

Raw Materials

Required raw materials are ammonia, phosphoric acid and potassium chloride. Urea is required to increase the nitrogen content where necessary. Fillers (sand, dolomite, etc.) and coating agents (clay, soapstone, etc.) are also required for certain grades.

Methods of Manufacture

Ammonia and phosphoric acid (48-54% P_2O_5) in the required proportion are metered to the preneutralizer and the resultant slurry of ammonium phosphate is pumped in a granulator for distribution over a bed of recycled material. The granulator may be a blunger or a rotating drum. Ammoniation is continued in the granulator to achieve the desired N : P ratio. Solid materials like urea, potash and filler are also added to make up the required formulation. The granules discharged from the granulator are dried, screened, cooled and coated with a coating agent (clay or powdered soapstone) to improve the storage properties. Flow diagram of NPK complex fertilizer production is shown in Fig. 8.3. Recently, pipe reactors have replaced preneutralizers as they are more energy efficient and simple to operate.

Product Characteristics

Physical characteristics: Granular NPK complex fertilizers are free-flowing with specific gravity of 1.54 and bulk density of 865 kg/m^3. Their CRH at 37°C is 60-65%. They can be stored under normal warehousing conditions. Moist conditions, however, promote caking.

Chemical characteristics: NPK complex fertilizers have generally less nitrogen than P and K. The N content varies from 10 to 22% which may be in partly ammoniacal and partly urea form. Total P_2O_5 content varies from 10 to 35% which is partly water-soluble and partly citrate-soluble. Potash (K_2O) content varies from 0 to 26% which is water-soluble.

MIXED FERTILIZERS

Mixed fertilizers are made by physical mixing of two or more fertilizers, containing two or more major plant nutrients, with or without some inert material. They may be mixture of NPK fertilizers, mixture of micronutrients or a mixture of NPK fertilizers with micronutrient fertilizers. The product may be in pulverized form or in granular form.

Mixed fertilizers serve the same purpose as complex fertilizers as they contain two or more major plant nutrients. They contain essential plant nutrients in proportions tailor-made to suit local conditions and ready for application. This saves the farmers' time in purchase, storage, handling and applying several individual fertilizers.

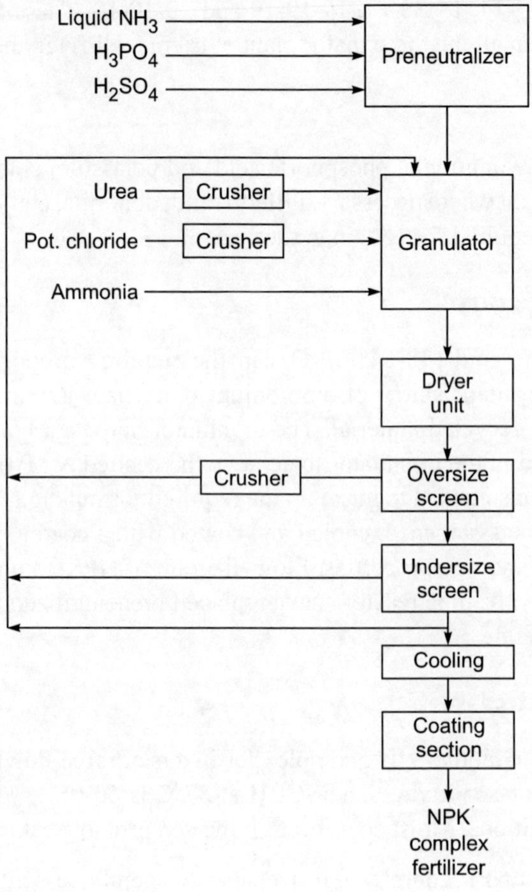

Fig. 8.3 Flow diagram of NPK complex fertilizer production

Choice of Fertilizer Grade

Fertilizer grade refers to the minimum guarantee of the plant nutrients in terms of total N, P_2O_5 and K_2O expressed as percentage in that order. For example, mixed fertilizer of 10-20-20 grade will contain a minimum of 10% N, 20%, P_2O_5 and 20% K_2O.

Fertilizer ratio refers to the ratio of the percentage of N, P_2O_5 and K_2O in the fertilizer. For example, the fertilizer grade 10-20-20 has fertilizer ratio of 1 : 2 : 2.

The mixed fertilizers or fertilizer mixtures are tailor-made to suit the local soil and crop requirements. The grade of the fertilizer mixtures largely determines the suitability of the product for a given crop growing in a particular soil condition. Generally, high analysis fertilizer mixtures (greater than 30% N + P_2O_5 + K_2O) are preferable for economic and profitable production. The advantages of high analysis fertilizers include lower cost per unit of plant nutrient, lower transportation cost, less storage space requirement, and less labour requirement for handling. The choice of the grade of the fertilizer mixture is determined by three factors: soil fertility status, crops grown in the area, and availability of basic materials.

Soil fertility status: The information about the fertility status of soil in a locality or district or state is available from the Soil Testing Laboratory serving that locality. This information is summarized in the form of soil test summaries and indicates the overall sufficiency or deficiency of major nutrients. Special attention should be paid to P and K status of the soils because these nutrients are applied as basal dressing whereas the extra amount of N can be added by the farmers through straight N fertilizers.

The information about the deficiency of sulphur and micronutrients is also available from the Soil Testing Laboratory.

Crops grown in the area: Crops differ in their nutrient requirement. For example, leguminous crops need more phosphorus than cereals. Similarly, the crops like potato and sugar cane need more potassium than other crops. Therefore, the fertilizer mixture for potato and sugar cane should contain sufficient K, particularly for potassium-deficient soils.

Availability of basic materials: In general, locally available basic materials are used for making fertilizer mixtures. If only low analysis basic materials are available, it is not possible to produce high analysis fertilizer mixture. For instance, if only ammonium sulphate and single superphosphate are available, it will not be possible to get a fertilizer mixture containing more than 20% plant nutrients.

Selection of Ingredients

The basic materials which are used as ingredients of mixed fertilizers determine the grade and quality of final product. The ingredients should therefore be selected carefully keeping in mind the quality of the final product. However, sometimes a compromise is necessary when the desired high quality ingredient is not available locally or within accessible limits.

The following properties of the material determine its eligibility as a basic material for mixed fertilizers.

1. *Nutrient content:* This is perhaps the most important criterion for the selection of a basic material for mixed fertilizer. For production of high analysis mixed fertilizer high analysis basic materials are required.
2. *Chemical form of nutrient:* Nitrogen may be present in ammoniacal, nitrate or amide form. Generally, nitrate form of nitrogen should be low in the final product; hence, ammoniacal form is preferred, which may be obtained at the lowest cost from anhydrous ammonia. Urea may also be used as basic material. In P source the phosphorus should mostly be in water-soluble form so as to ensure crop response to the mixed fertilizer. Materials like rock phosphate, though inexpensive, should, therefore not be used as ingredient unless the mixed fertilizer is to be used in highly acid soils only.
3. *Content of other essential nutrients*: It is desirable to use such basic materials, if available, which provide secondary or micronutrients in addition to major nutrients (NPK). For example, ammonium sulphate provides sulphur in addition to N while single superphosphate provides sulphur and calcium in addition to P.

4. *Absence of toxic elements*: The basic ingredients should be free from toxic elements which may be injurious to plants or human health. For this reason, use of phosphogypsum or fly ash as filler in mixed fertilizers is not desired.
5. *Hygroscopicity:* The hygroscopic materials absorb moisture quickly from the air when relative humidity exceeds certain limit. This makes the mixtures sticking and susceptible to caking during storage. This property of fertilizers or basic materials is expressed as Critical Relative Humidity (CRH). Use of hygroscopic materials like calcium nitrate and ammonium nitrate is therefore avoided.
6. *Particle size:* Different fertilizer materials are available in different forms such as powder, crystal, granule and prills. Because of their different particle size and bulk density, mixing of such ingredients may result in segregation.
7. *Compatibility with other ingredients*: Besides the problem of segregation due to dissimilar particle size of ingredients, mixing of some ingredients may result in loss of nutrients or increased hygroscopicity. Hence, judicious selection of the ingredients becomes necessary.

A chart indicating the guidelines for mixing fertilizer materials is given in Fig. 8.4.

In mixing straight fertilizers, the following points should be kept in view:

(i) Fertilizers containing ammonia or urea should not be mixed with lime or basic reactive fertilizers like basic slag.

(ii) Water-soluble P fertilizers, e.g. single superphosphate, triple superphosphate, ammonium phosphates should not be mixed with those fertilizers which contain free lime because this may convert a portion of the soluble phosphate into an insoluble form.

(iii) Easily soluble and hygroscopic fertilizers like urea, calcium ammonium nitrate and potassium salts tend to cake or form lumps after mixing. Such fertilizers should be mixed shortly before use.

In addition to the physical and chemical properties of the ingredients, an important consideration is its assured availability in sufficient quantity for the operation of the mixed fertilizer plant. Commonly used ingredients are:

N sources: Ammonia, ammonium nitrate, urea.

P sources: Single superphosphate, triple superphosphate, monoammonium phosphate (MAP), DAP, phosphoric acid.

K sources: Muriate of potash.

Micronutrient sources: Borax, ferrous sulphate, zinc sulphate, copper sulphate, manganese sulphate.

Conditioners (coating agents): Soapstone and clay are used as coating agents for granular mixed fertilizers to give good drilling condition, and to reduce caking during storage.

Conditioners (parting agents): Rice husk and groundnut shell are used as conditioners or parting agents for pulverized mixed fertilizers to give good drilling condition, and to reduce caking during storage.

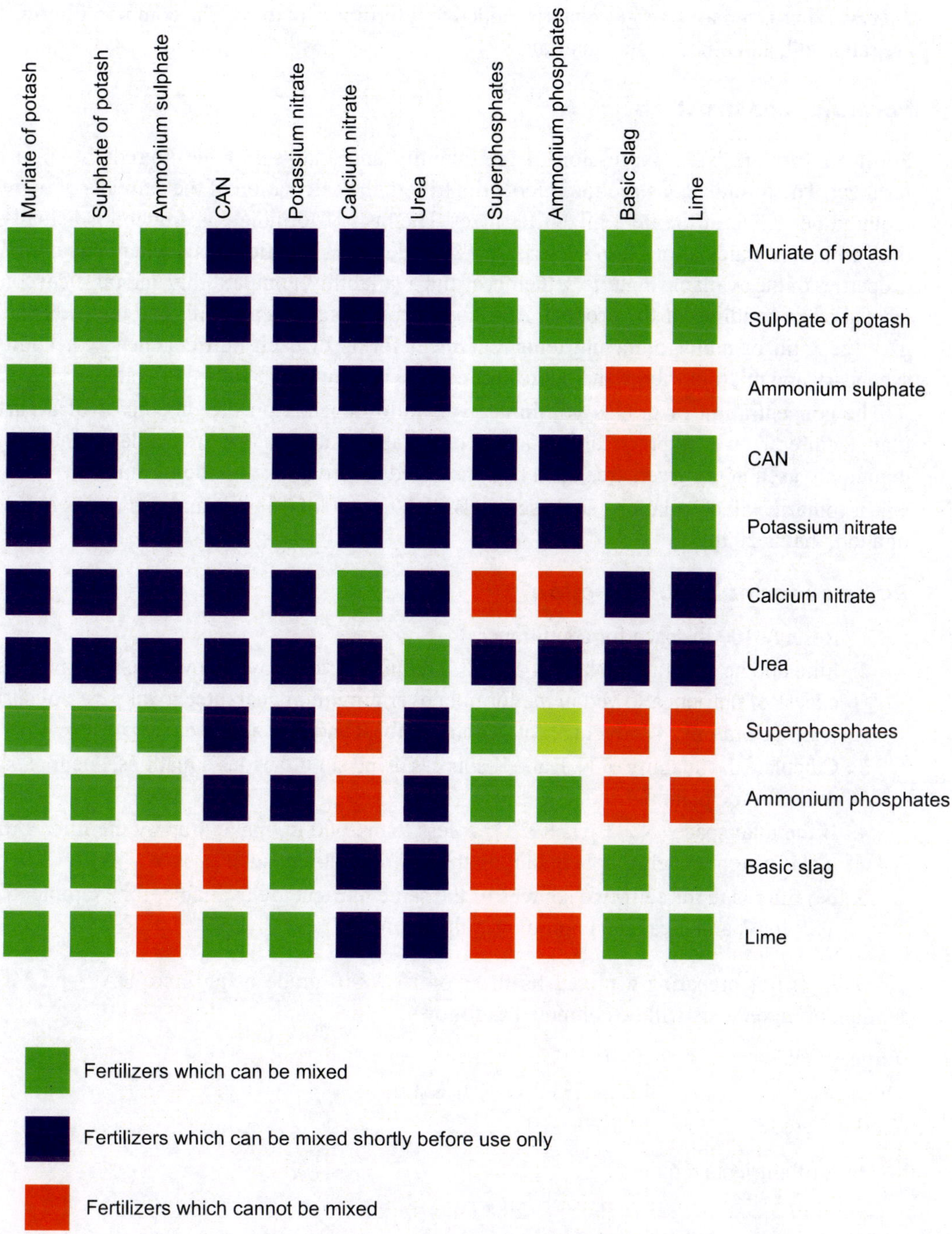

Fig. 8.4 Guidelines for mixing fertilizers

Fillers: Filler is a make-weight material added to a fertilizer mixture. The common filler materials are sand, soil, and other waste materials.

Formula Construction

Fertilizer formula is an expression of the quantity and analysis of the ingredients in a mixed fertilizer. The formula construction refers primarily to the calculation of the amounts of ingredients required per tonne of the final mixed fertilizer. The task of formula construction is actually more than arithmetic calculation. It is necessary to give due consideration to the chemical and physical properties of the available materials, their cost and availability, compatibility and physical condition in relation to handling of the product. Agronomic considerations pertaining to the soils and crops give the ratio of major plant nutrients, chemical forms of each nutrient, need for addition of secondary and micronutrients and degree of residual acidity which can be tolerated.

The concentration of grades within a given ratio depends on the objectives of a particular manufacturer. However, practical limitations of suitable materials that are available and the grades commonly used in the given area must be considered before finalizing the formula.

Customarily, all calculations are based on British ton (2000 lb) basis and one unit is equal to 1% of a ton, that is 20 lb.

Rules of formula construction

1. Reserve 100 lb space for conditioner.
2. Find out the quantity of N, P_2O_5 and K_2O required after allowing overage. The overage is the excess of nutrient allowed in the formula over minimum guarantee to take care of shrinkage loss. Commonly, 4% overage for N and 2% for P and K is allowed.
3. Calculate the quantity of N, P and K sources using simultaneous equations, one for space and other for the plant nutrient.
4. If the total space (N + P_2O_5 + K_2O) is less than 2000 lb, make it up by the filler. Filler is a make-weight material not containing the desired nutrient such as sand.
5. No filler is required if two sources of the same nutrient are available, for example, SSP and TSP for P, and urea and ammonium sulphate for N.

Example 1: For preparing a mixed fertilizer of 20-20-10 grade using urea, DAP and MOP, the amount of ingredients will be computed as follows:

Target grade	20-20-10
Overage	4% N, 2% P_2O_5, 2% K_2O
Working grade	20.8-20.4-10.2

Units of nutrients required

N	20.8 units	or	20.8×20 lb
P_2O_5	20.4 units	or	20.4×20 lb
K_2O	10.2 units	or	10.2×20 lb

Amount of DAP (for P)	$20.4 \times 20 \times 0.46 = 887$ lb
Amount of N available from DAP	$887 \times 0.18 = 160$ lb or 8 units
Amount of N required from urea	$20.8 - 8 = 12.8$ units
Amount of urea required	$12.8 \times 20 \times 0.46 = 557$ lb
Amount of MOP required	$10.2 \times 20 \times 0.60 = 340$ lb
Amount of conditioner	Adjusted in filler
Amount of filler required	$2000 - (887 + 557 + 340) = 216$ lb

Example 2: For preparing a mixed fertilizer of 15-15-15 grade using urea, SSP, TSP and MOP, the amount of ingredients will be computed as follows:

Target grade	15-15-15
Overage	4% N, 2% P_2O_5, 2% K_2O
Working grade	15.6-15.3-15.3

Units of nutrients required

N	15.6 units	or	15.6×20 lb
P_2O_5	15.3 units	or	15.3×20 lb
K_2O	15.3 units	or	15.3×20 lb

Amount of urea required	$15.6 \times 20 \times 0.46 = 678$ lb
Amount of MOP required	$15.3 \times 20 \times 0.60 = 510$ lb
Amount of conditioner	100 lb
Space remaining	$2000 - (678 + 510 + 100) = 712$ lb
Amount of P_2O_5 required	$15.3 \times 20 = 306$ lb

Let amount of SSP = a lb

and amount of TSP = b lb

$$a + b = 712 \quad \text{(i) Space equation}$$
$$0.16a + 0.46b = 306 \quad \text{(ii) Nutrient equation}$$

Multiplying equation (i) by 0.46

$$0.46a + 0.46b = 327.5 \quad \text{(iii)}$$
$$(-)\ 0.16a + 0.46b = 306 \quad \text{(ii)}$$
$$0.30a = 21.5$$
$$a = 21.5/0.3 = 72 \text{ lb}$$
$$b = 712–72 = 640 \text{ lb}$$

Amount of SSP required = 72 lb

Amount of TSP required = 640 lb

Amount of filler required = Nil

Method of Manufacture

Mixed fertilizer plants are characterized by the process used and product produced. There are four general types of processes and products:

(a) **Pulverized Mixed Fertilizers**

It is simpler to make pulverized mixture as it requires very little equipment and is formulated from manufactured materials. Ammoniation is often used for economy of N source and good physical condition of the final product. The ammoniation rates are 6 and 4 lb ammonia per unit (20 lb) P_2O_5 for single- and triple superphosphate, respectively. Principal ingredients are SSP, TSP and MOP. Parting agents are used to ensure physical stability. Four steps are involved in production of pulverized mixed fertilizers:

(i) Screening and grinding solid ingredients
(ii) Ammoniation of superphosphate
(iii) Mixing in a rotary mixer
(iv) Curing for 2-4 weeks

If ammoniation of superphosphate is not done, the process becomes much simpler. The ingredients are weighed and charged in a rotary or continuous mixer, and the mixed product is bagged. The product, however, suffers from certain disadvantages like dustiness, segregation during handling and lack of free-flowing nature, which causes difficulty in its uniform application through a seed drill.

(b) **Granulated mixed fertilizers**

Granulated mixed fertilizers are made by intimately mixing two or more fertilizers with or without inert material, and granulating them together. The granulation eliminates the limitations of pulverized mixed products. The granulated mixed fertilizers are dry, dust-free, free-flowing and easy to apply. There is no segregation of particles and the product is less susceptible to caking during storage.

The ingredients are fed in a weighing hopper in proportionate quantities, ground and then mixed in a rotary drum mixer and granulated in a rotary drum granulator where a small quantity of water is added. The granules containing about 10% moisture are fed to the dryer, cooled and screened.

(c) **Bulk blending**

Bulk blending technology is simple as it involves physical mixing of two or more granular products having similar particle size. Urea, DAP and granular MOP are principal materials used for bulk blending. The process has advantages of economy of mixing the finished products, versatility of grade or nutrient ratio and convenience of delivering the bulk blended materials directly to the field. However, there are certain limitations such as risk of segregation of the materials if the particle sizes of the ingredients do not match. Also, incompatible materials like urea and SSP cannot be blended because of poor physical quality of the final product.

(d) **Fluid mixed fertilizers**

In developed countries fluid fertilizers are very popular as they are easy to transport and handle, provide uniform application of plant nutrients, allow mixing and uniform application of micronutrients, herbicides and insecticides. There are three main categories of fluid fertilizers:

1. **Clear liquids:** These are true solutions with a limitation of low analysis (8-8-8 or 5-10-10) because of the risk of salting out or crystallization at low temperatures.
2. **Suspension fertilizers:** These are mixtures of liquids and finely divided solids in which the solids do not settle rapidly and can be redispersed readily by agitation. Certain types of clay are usually added as suspending agents. Suspension fertilizers can be produced to give the same concentration of plant nutrients as solid fertilizers.
3. **Slurry fertilizers:** These are mixtures of liquids and finely divided solids in which the solids settle rapidly in the absence of agitation and form a firm layer at the bottom of the tank.

Nitrogen in the fluid fertilizers is usually in the form of ammonium nitrate or urea along with diammonium phosphate and monoammonium phosphate. Phosphorus in the clear liquid fertilizer is 100% water-soluble. Muriate of potash is the most common source of potassium in all fluid fertilizers.

When metal micronutrients (zinc, iron, manganese and copper) are added to clear liquid fertilizers, precipitation often occurs as a result of the reaction with phosphates. Chelates of metal micronutrients can be mixed with liquid fertilizers without the risk of causing precipitation but they are more expensive than inorganic metal salts. Soluble boron source can be added to liquid fertilizers.

Suspension or slurry fertilizers may contain the same source of micronutrients as those found in dry solid fertilizers. Their precipitation does not pose any problem as they still remain in suspension and can be distributed through the mixed fertilizer.

Selected References

FAI (2001). *Handbook on Fertilizer Technology*. 6th ed., Fertiliser Association of India, New Delhi. pp. 273.

Mordtvet, J.J., Murphy, L.S. and Follett, R.F.H. (1999). *Fertilizer Technology and Application*. Meister Publishing Company, Euclid Ave., Willoughby, OH.

UNIDO/IFDC, Ed. (1998). *Fertilizer Manual*, 3rd ed., Kluwer Academic Publishers, Dordrecht, the Netherlands.

CHAPTER

9

Secondary Nutrient Fertilizers

Sulphur, calcium and magnesium are macronutrients required in relatively large quantities for good crop growth. They are collectively known as 'secondary nutrients' because they are generally not deficient in soils and they are ingredients of soil amendments such as liming materials and gypsum. But this notion about the deficiency of these plant nutrients has changed recently in view of widespread deficiency, especially of sulphur throughout the world. The crops require as much sulphur and magnesium as phosphorus whereas for many crops the calcium requirement is greater than that for phosphorus.

SULPHUR FERTILIZERS

Recent upsurge in sulphur deficiency is attributed to the following:

1. Lack of incidental application of sulphur through sulphur-containing fertilizers like ammonium sulphate and single superphosphate.
2. Increased use of high analysis sulphur-free fertilizers like urea, DAP.
3. Decreased use of farm manures such as FYM and compost which supply good amount of sulphur and micronutrients.
4. Intensive cropping without use of sulphur-containing fertilizers.
5. Decreased atmospheric deposition of sulphur due to low or no emission loss of sulphur gases from the factories as pollution control measure.

Need for sulphur fertilizers will increase in future as the soils today considered adequate in sulphur may become depleted rapidly because of adoption of multiple cropping with use of sulphur-free fertilizers and improved management to boost crop yields that will accelerate the withdrawal rate of sulphur from soils.

Classification of Sulphur Fertilizers

On the basis of the chemical form of sulphur present in the product, sulphur-containing fertilizers may be grouped into three broad groups:

1. Fertilizers containing elemental sulphur, e.g. elemental sulphur, sulphur-bentonite, urea-sulphur, ammonium phosphate-sulphur, etc.
2. Fertilizers containing sulphate, e.g. ammonium sulphate, superphosphates, gypsum, potassium sulphate, etc.
3. Fertilizers containing other forms of sulphur, e.g. ammonium thiosulphate, ammonium bisulphite, ammonium polysulphide, pyrites.

Table 9.1 Sulphur-containing fertilizers

Material	*Plant nutrient content (%)*				
	N	P_2O_5	K_2O	S	Others
Ammonium bisulphite	14.1	–	–	32.3	
Ammonium sulphate	21	–	–	24.2	
Ammonium phosphate-sulphate	20	20	–	13–15	
Ammonium phosphate-sulphur	16	38	–	12–15	
Ammonium thiosulphate	18.9	–	–	43.3	
Ammonium polysulphide	20.5	–	–	45	
Ammonium polysulphide solution	20	–	–	40	
Ferrous sulphate	–	–		18.8	32.8 (Fe)
Gypsum (Agricultural grade)	–	–	–	13	17 (Ca)
Magnesium sulphate	–	–	–	13	9.8 (Mg)
Potassium sulphate	–	–	50	17.6	
Potassium-magnesium sulphate	–	–	22	22	11 (Mg)
Pyrites (Agricultural grade)	–	–	–	20–25	14–18 (Fe)
Sulphur	–	–	–	100	
Sulphur (granular with additives)	0–7	–	–	65–95	
Single superphosphate	–	16		13	
Triple superphosphate	–	46	–	1.5	
Urea-sulphur	36–40	–	–	10–20	
Zinc sulphate	–	–	–	17.8	36.4 (Zn)

Fertilizers Containing Elemental Sulphur

Elemental Sulphur: Sulphur is a solid, light yellow non-metallic element. Elemental sulphur is commonly used in agriculture as fungicide and as soil amendment for alkali lands. In industry, elemental sulphur is needed for production of sulphuric acid which is required for production of a number of products including fertilizers.

Recently, sulphur contamination in crude oil has provided a major source of sulphur. In 1990, mined sulphur made up for around a third of the world's total sulphur supply; it now makes up less than one per cent, with sulphur extracted from crude oil and natural gas making up for around 98% of the world's total sulphur production.

The methods of sulphur mining and recovery from sour gas are described in Chapter 2.

Product Characteristics

Elemental sulphur is light yellow, inert crystalline solid. It is odourless and insoluble in water. Under certain conditions, sulphur dust and air form a potentially explosive mixture. The particle size of elemental sulphur for agricultural use is usually fine (100% should pass through 16 mesh and 50% through 100 mesh screen). The finer the particle size, the greater the surface area and faster the oxidation of sulphur to release sulphate ions which are available to plants.

Sulphur is insoluble in water and oxidizes slowly on contact with soil, air or the atmosphere. Sulphur ignites at 250°C. When burnt it produces sulphur dioxide, a pungent gas, which is widely used in the manufacture of sulphuric acid. About 75-85% of elemental sulphur is used for this purpose only.

Sulphur-Bentonite: For agricultural use, elemental sulphur is also marketed as prilled sulphur which is 90% sulphur and 10% bentonite (swelling clay). Prilled sulphur is dust-free and safe to handle with reduced risk of explosion hazard.

Urea-Sulphur: It contains 40% nitrogen and 10% sulphur. Urea-sulphur is produced by fusing urea and elemental sulphur, followed by prilling. The product has excellent storage and handling properties.

Sulphur-coated urea containing 35% nitrogen and 19% sulphur has recently shown promise as slow-release nitrogen fertilizer. Its production has been discussed in the chapter on nitrogen fertilizers.

Ammonium phosphate-Sulphur: This product contains 12-15% sulphur which is obtained by addition of sulphur to DAP during granulation step.

Fertilizers Containing Sulphate

Production and characteristics of ammonium sulphate, single superphosphate and ammonium phosphate sulphate are discussed in the chapters on nitrogen, phosphate and complex fertilizers. Ammonium sulphate is nitrogen fertilizer which contains more sulphur (24%) than nitrogen (20.5%). Likewise, single superphosphate contains about 12% sulphur, in addition to 16% P_2O_5. Ammonium phosphate sulphate is complex fertilizer which contains 20% N, 20% P_2O_5 and 13-15% sulphur. In these fertilizers, sulphur is present in sulphate form but there is difference in solubility. Ammonium sulphate and ammonium phosphate sulphate are readily soluble in water, whereas single superphosphate contains sulphur in the form of gypsum (calcium sulphate) which is slowly soluble but meets the crop requirements.

Sulphate of potash and sulphate of potash-magnesia are also good sources of potash as well as sulphur.

Gypsum: Gypsum is a naturally occurring mineral which is commonly used as soil amendment (source of calcium) for sodic (alkali) soils. It is also used as source of plant nutrient sulphur for all types of soils.

Gypsum is used in the cement industries. Naturally occurring crystalline form of high purity gypsum is known as *selenite,* which is used extensively in the ceramics industries and to manufacture surgical grade plaster of Paris.

Method of Extraction

Natural deposits of gypsum are found in arid areas of the world. In India, there are large deposits of gypsum located in many districts of Rajasthan, Gujarat and south India. The total reserve is estimated at more than 70 million tonnes. Gypsum is mined by open cast method. The material contains 50-90% gypsum. Agricultural grade gypsum is produced by grinding the gypsum ore and screening.

By-product gypsum, also known as phosphogypsum, is available from wet process phosphoric acid plants, which can be used as source of sulphate and calcium in agriculture and industry.

Product Characteristics

Pure gypsum is white, powdery material but agricultural grade gypsum is a greyish powder. It is an inert and odourless solid.

Pure gypsum has the formula $CaSO_4.2H_2O$ which contains 18.6% sulphur and 23.2% calcium. The agricultural grade natural gypsum contains 80-85% gypsum. It is sparingly soluble in water (2 g per litre). It is the more soluble form of calcium than limestone. Unlike elemental sulphur or ammonium sulphate it will not produce acidity, though it will lower the pH of sodic soils by replacing sodium from the soil colloids by calcium.

Fertilizers Containing Other Forms of Sulphur

In this category, there are four fertilizer materials which contain sulphur in reduced or partially oxidized form.

Ammonium thiosulphate: This is perhaps the most widely used sulphur source in the liquid fertilizer industry. It is prepared by reacting sulphur dioxide and aqueous ammonia to form at first an ammonium sulphate solution which reacts further with elemental sulphur to form ammonium thiosulphate, $(NH_4)_2S_2O_3$. The product is a solution containing 12% nitrogen and 26% sulphur.

Ammonium bisulphite: This is a by-product of smelting operations and is available commercially as a 55-60% solution of ammonium bisulphite (NH_4HSO_3) containing 8.5% nitrogen and 17% sulphur. Small amounts of ammonium sulphite and ammonium sulphate are usually present. This solution has no pressure, a pH of 5.2, and crystallization temperature of 0°C (32°F).

Ammonium polysulphide: This is a reddish-brown to black coloured solution having a strong odour of hydrogen sulphide (like rotten egg). It contains 20% nitrogen and 40-45% sulphur. Ammonium polysulphide, $(NH_4)_2S_{x,}$ is obtained by reacting aqua ammonia with hydrogen sulphide in the presence of heat and agitation. Small amounts of sulphate, thiosulphate and elemental sulphur may be present in the final product.

The product is strongly alkaline having a pH of 10 or more. This is a quite reactive material and is not compatible with acidic solutions since it decomposes in an acid medium, liberating hydrogen sulphide and forming colloidal sulphur.

Iron pyrites: It is also known as 'fool's gold' due to yellow colour similar to gold. Pure pyrites is chemically iron disulphide (FeS_2) which contains 46% iron and 53% sulphur. Pyrites is mined for sulphur rather than for iron for obvious reason that better iron ores are available. Deposits of iron pyrites are found in many countries in Asia, Europe and North America and also in Australia. About 98 million tonnes reserve of pyrites is found in India at Amjhore in Rohtas district of Bihar; Saladipura in Rajasthan; and in Andhra Pradesh and Karnataka (Table 9.2). However, most of the ores in these deposits are of poor grade. The high grade ore containing 22-25% sulphur is used for production of sulphuric acid (Table 9.3) while the low grade pyrites (15-20% sulphur) is marketed as soil amendment.

Table 9.2 Reserves of pyrites in India (1995)

State	*Total reserves (million tonnes)*	*Grade*
Andhra Pradesh	0.7	Beneficiable grade
Karnataka	10.87	Soil amendment
Bihar	52.0	Soil amendment and unclassified
Rajasthan	34.35	Beneficiable grade
Total	97.92	

Table 9.3 Chemical composition of high grade Amjhore pyrites and Saladipura pyrites

Constituent	*Amjhore pyrites (%)*	*Saladipura pyrites (%)*
Sulphur	22.6	23.6
Iron	19.03	26.5
Carbonaceous matter	2.56	–
Silica	46.5	25.7
Aluminium	9.0	10.9
MgO	0.1	1.73
CaO	–	7.87
K_2O	–	0.96
Na_2O	–	0.41
Cu	Trace	100 ppm
Zn	–	100 ppm
Moisture	3-5	–

Direct application of pyrites as nutrient source has seldom been considered. Some recent work done in Australia has shown that the rate of oxidation of pyrite in soil is very slow which limits its effectiveness as sulphur source. Particle size distribution of the material has a tremendous effect on its oxidation rate. Finer the particle size, faster the rate of oxidation and greater the effectiveness.

CALCIUM FERTILIZERS

Calcium is present in several fertilizers supplying phosphorus, particularly, single superphosphate and triple superphosphate which contain 18-21% and 12-14% calcium, respectively. CAN also contains about 20% lime (8% calcium). Although calcium is not usually formulated into mixed fertilizers, it is present as component of the ingredients used for production of mixed fertilizers.

Liming materials such as limestone and dolomite, which are applied to acid soils to neutralize the acidity, are the primary sources of calcium. In the situations where calcium is required without the need for correcting soil acidity, gypsum is used.

Gypsum is widely used as soil amendment on sodic (alkali) soils in arid and semi-arid climates. The calcium of gypsum replaces sodium on the exchange complex and then sodium is leached out of root zone resulting in reduction of sodium saturation and soil improvement.

Characteristics of some common calcium sources used in fertilizer and liming materials are given in Table 9.4.

Table 9.4 Characteristics of some common calcium sources used in fertilizer and liming materials

Source	*Common name*	*Chemical formula*	*Calcium, %*	*Water solubility* (g/L at 25°C)
Burned lime	Quick lime	CaO	70	1.2
Hydrated lime	Slaked lime	$Ca(OH)_2$	50	1.6
Calcitic lime	Limestone	$CaCO_3$	36	0.1
Dolomitic lime	Dolomite	$CaCO_3.MgCO_3$	17	> 0.1
Basic slag	Basic slag	$(CaO)_5.P_2O_5.SiO_3$	29	> 0.1
Gypsum	Gypsum	$CaSO_4.2H_2O$	22	2.4
Single superphosphate	Single superphosphate	$Ca\ (H_2PO_4)_2.CaSO_4$	20	10.0
Triple superphosphate	Triple superphosphate	$Ca\ (H_2PO_4)_2$	13	18.0
Rock phosphate	Rock phosphate	$Ca_{10}(PO_4)_6.F_2$	33	0.02
Calcium ammonium nitrate	Calcium ammonium nitrate	$NH_4NO_3.CaCO_3$	8	0.1

MAGNESIUM FERTILIZERS

In contrast to calcium, major nutrient fertilizers do not contain magnesium (Mg) except potassium-magnesium sulphate. Dolomite is commonly applied to the soils low in magnesium. Potassium-magnesium sulphate and magnesium sulphate (Epsom salt) are the most widely used materials in dry fertilizer formulations. Other materials containing magnesium are magnesia, MgO (55% Mg), magnesium nitrate, $Mg(NO_3)_2$ (16% Mg), magnesium chloride, $MgCl_2.10H_2O$ (8-9% Mg) and synthetic chelates (2-4% Mg). Characteristics of some common magnesium sources used in fertilizers are summarized in Table 9.5.

Table 9.5 Characteristics of some common magnesium sources used in fertilizers

Source	*Common name*	*Chemical formula*	*Magnesium* (%)	*Water solubility* (g/L at 25°C)
Magnesium oxide	Magnesia	MgO	45.0	0.0062
Magnesium chloride	Magnesium chloride	$MgCl_2.10H_2O$	8.7	Soluble
Magnesium nitrate	Magnesium nitrate	$Mg(NO_3)_2$	16.2	Soluble
Dolomitic lime	Dolomite	$CaCO_3.MgCO_3$	12	0.32
Kieserite	Kieserite	$MgSO_4.H_2O$	18.2	684
Langbeinite	K-Mag	$K_2SO_4.2MgSO_4$	11.2	Soluble
Magnesium sulphate	Epsom salt	$MgSO_4.7H_2O$	10.5	910
Magnesium chelates	Magnesium chelates		2–4	Soluble

Epsom salt, magnesium chloride, magnesium nitrate and synthetic chelates are well suited for application in clear liquids and foliar sprays. Potassium-magnesium sulphate is most widely used source of magnesium in suspension fertilizers.

Selected References

Havlin, J.L., Beaton, J.D., Tisdale, S.L. and Nelson, W.L. (2005). *Soil Fertility and Fertilizers*, 7th ed. Pearson Education, Upper Saddle River, N.J.

Mordtvet, J.J., Murphy, L.S. and Follett, R.F.H. (1999). *Fertilizer Technology and Application*. Meister Publishing Company, Euclid Ave., Willoughby, OH.

UNIDO/IFDC, (Ed.) (1998). *Fertilizer Manual*, 3rd ed., Kluwer Academic Publishers, Dordrecht, the Netherlands.

CHAPTER

10

Micronutrient Fertilizers

Seven elements essential for plant growth are commonly referred to as micronutrients because they are required by plants in small quantities compared to macronutrients. They are boron, copper, iron, manganese, molybdenum, zinc and chlorine. Except the last one, all micronutrients are required in concentrations as small as less than one half parts per million in the soil solution. Chlorine is found in plants in concentrations higher than that of phosphorus (P); however, it is considered as essential micronutrient. During past three decades, the area of deficiency of these plant nutrients has been increasing throughout the world. This increase in micronutrient deficient areas is attributed to the following:

1. Increased use of high analysis, micronutrients-free fertilizer like urea, DAP.
2. Decreased use of farm manures such as FYM and compost which supply good amount of micronutrients.
3. Intensive cropping without use of micronutrients-containing fertilizers.
4. Increased awareness about micronutrients deficiencies in order to identify the limiting factors in crop production.

Need for micronutrients fertilizers is likely to increase in future as the soils today considered adequate in micronutrients may become depleted gradually because of adoption of multiple cropping without use of micronutrients fertilizers and improved management to boost crop yields that will accelerate the withdrawal rate of micronutrients from soils.

CLASSIFICATION OF MICRONUTRIENTS FERTILIZERS

On the basis of the chemical form of the nutrients in the product, micronutrients fertilizers may be grouped into four broad classes:

- inorganic products
- synthetic chelates

- natural organic complexes
- fritted glass products (frits)

Inorganic Sources

Inorganic sources of micronutrients include oxides and carbonates, and metallic salts such as sulphates, chlorides and nitrates. The sulphates are the most common of the metallic salts and are sold in crystalline or granular form. Oxides of manganese and zinc are also commonly used and are sold as fine powders and in granular form. Because oxides such as ZnO and MnO are water-insoluble, their immediate effectiveness for crops is rather low in granular form. Also, the available divalent form of manganese in MnO will oxidize to the unavailable tetravalent form of manganese, so there is very little residual availability of manganese fertilizers for succeeding crops. Thus, agronomic effectiveness of granular MnO may be rather low. Since manganese in MnO_2 is in the unavailable form, it should not be used as a manganese fertilizer.

Oxysulphates are oxides, usually industrial by-products, which have been partially acidulated with sulphuric acid, and are generally sold in granular form. The percentage of water-soluble manganese or zinc in oxysulphates is directly related to the degree of acidulation by sulphuric acid. Research results have shown that about 35 to 50 per cent of the total zinc in granular zinc-oxysulphate should be in water-soluble form to be immediately effective for crops. Similar results would be expected for manganese-oxysulphate. Inorganic sources usually are the least costly sources per unit of micronutrient, but they may not always be the most effective for crops.

Synthetic Chelates

The word "chelate" is derived from the Greek word meaning 'claw'. In fertilizer technology, it refers to inorganic nutrients that are enclosed by an organic molecule. It has a ring structure produced when a metal ion combines with two or more electron donor groups to form a single molecule. Metals bound in chelate rings lose their ability to act as ions and therefore are unable to take part in chemical reactions that precipitate them or render them unavailable to plants. The heavy metals essential for plant growth such as iron, manganese, copper and zinc tend to become less available to plants as the soil pH rises above 6.3 due to formation of insoluble oxides or hydroxides.

The deficiencies of these elements occur in the neutral and alkaline soils having pH above 6.3. The metal chelates offer a means of minimizing the chemical reactions of heavy metals ions in the soil. These sources are formed by combining a chelating agent with a metal through coordinate bonding. All polyvalent cations will form chelates but their stability differs from one another. The stability of the metal-chelates in decreasing order is:

$$Fe^{3+} > Cu^{2+} > Zn^{2+} > Fe^{2+} > Mn^{2+} > Ca^{2+} > Mg^{2+}$$

Ferric iron chelate is more stable than any of other chelates of the metal elements that are essential for plant growth. Different kinds of chelating agents are used to produce metal chelates. Some are effective on acid soils, some on calcareous soils (containing free lime) and some are effective on both acid and alkaline soils. Commonly used chelating agents are:

1. EDTA (Ethylene Diamine Tetraacetic Acid)
2. DTPA (Diethylene Triamine Pentaacetic Acid)
3. HEDTA (Hydoxyethyl Ethylene Diamine Triacetic Acid)
4. CDTA (Cyclohexane Diamine Tetraacetic Acid)
5. EDDHA (Ethylene Diamine Di-o-Hydroxyphenylacetic Acid)
6. NTA (Nitrilo Triacetic Acid)

Relative effectiveness for crops per unit of micronutrient as soil-applied chelates may be from two to five times greater than that of inorganic sources, while chelates costs per unit of micronutrient may be five to 100 times higher.

Chelated nutrients are compatible with dry fertilizers mixtures (2.5 to 10 kg of iron chelate per tonne has proved effective). However, they cannot be mixed with liquid fertilizers that contain ammonia because the higher pH precipitates the metals from the chelating agent.

Natural Organic Complexes

These complexes are made by reacting metallic salts with some organic by-products of the wood pulp industry or related industries. Some common types of these complexes are the lignosulphonates, polyflavonoids and phenols. The types of chemical bonding of the metals to the organic components are not well understood. Some bonds may be coordinate as in the chelates, but other types of chemical bonds also may be present. While natural organic complexes are less costly per unit of micronutrient, they usually are less effective than synthetic chelates. They are also more readily decomposed by microorganisms in soil. These sources are more suitable for foliar sprays and mixing with fluid fertilizers.

Frits

Fritted glassy products (frits) are brown-coloured pulverized or granular materials that slowly release micronutrients such as boron, copper, iron, manganese, molybdenum and zinc. Micronutrient concentrations vary from 2 to 25 per cent, and more than one micronutrient may be included in a fritted product. Fritted micronutrients generally are used only on sandy soils in regions of high rainfall where leaching occurs. This class of materials is more appropriate for micronutrient maintenance programmes than for correcting severe micronutrient deficiencies. Therefore, frits only have a small share of the micronutrient market.

For manufacture of frits, proper proportions of feldspar, soda ash, silica, fluorspar (CaF_2), cryolite ($AlF_3.3NaF$), sodium nitrate and borax are thoroughly mixed. The mix is fused in a smelting furnace at 900 to 1000°C. The molten material is quenched in water, dried and milled. The final product is finely ground, slightly soluble fritted glass. The solubility is controlled by particle size and changes in matrix composition. Typical composition of two products is given in Table 10.1.

Table 10.1 Typical composition of micronutrient frits

Ingredient	*Frit 504*	*Salt mix*
Iron	14.0	7.5
Manganese	7.0	8.2
Copper	7.0	3.2
Zinc	7.0	4.5
Boron	3.8	1.5
Molybdenum	0.07	8.05

Most frits are used as ingredient of mixed fertilizers and very little is added directly to the soil. Frits are not yet widely used.

IRON FERTILIZERS

The common sources of iron that are used as fertilizer are listed in Table 10.2. The most common iron salt used as fertilizer is ferrous sulphate which is also known as green vitriol. Ferrous ammonium sulphate is another soluble source of iron. The effectiveness of the soil application of soluble iron fertilizers for correcting iron deficiency in crops is poor because of rapid conversion of soluble iron sources into insoluble forms like ferric hydroxide. Their foliar application is therefore preferred for correction of iron deficiency.

Table 10.2 Common iron fertilizers

Fertilizer material	*Chemical formula*	*Per cent iron*	*Solubility*
Ferrous sulphate	$FeSO_4.7H_2O$	19	Soluble
Ferric sulphate	$Fe_2(SO_4).3H_2O$	23	Soluble
Ferrous oxide	FeO	77	Insoluble
Ferric oxide	Fe_2O_3	69	Insoluble
Ferrous ammonium sulphate	$FeSO_4.(NH_4)_2SO_4.6H_2O$	14	Soluble
Ferrous ammonium phosphate	$Fe(NH_4)PO_4.H_2O$	29	Slightly soluble
Iron chelates	NaFeEDTA	5-14	Soluble
	NaFeHEDTA	5-9	Soluble
	NaFeEDDHA	6	Soluble
	NaFeDTPA	10	Soluble
Natural organic materials		5-10	Slightly soluble

Methods of Manufacture

Ferrous sulphate is a by-product of pickling operation in steel manufacture. In a small-scale industry it is produced by treating iron scrap with sulphuric acid. In case of both the processes, the common chemical reaction may be expressed as

$$Fe + H_2SO_4 = FeSO_4 + H_2$$

For production of one tonne ferrous sulphate, about 310 kg iron scrap and 370 kg sulphuric acid are required. Ferric sulphate is made by treating flue dust, obtained from a Cottrell precipitator in smelting operations, with sulphuric acid.

Iron chelates include monosodium ferric ethylene diamine tetraacetate (NaFeEDTA), disodium ferric ethylene diamine tetraacetate (Na_2FeEDTA), monosodium ferric diethylene triamine pentaacetate (NaFeDTPA), monosodium ferric hydroxyl ethylene diamine triacetate (NaFeHEDTA). Among micronutrient metal-chelates, iron chelates are the most widely used chelates.

Product Characteristics

Ferrous sulphate is a blue-green cystalline salt whereas ferric sulphate is a brownish red salt. Ferrous ammonium sulphate is greenish white and ferrous ammonium phosphate is green salt. Ferrous iron salts are more soluble than ferric iron salts and also more available but some of them are unstable. For example, ferrous sulphate readily loses water and undergoes oxidation resulting in conversion to ferric sulphate with a change in colour from blue green to a brown. They are good sources of iron for foliar spray but less effective as soil application.

Iron chelates are water-soluble and most stable in the soil. Among iron chelates, Fe-EDDHA is the most stable chelate. The chelates are 4-5 times more effective as soil application than soluble salts but they are too expensive for soil application. For foliar application, iron chelates are nearly as effective as ferrous sulphate.

ZINC FERTILIZERS

The common sources of zinc that are used as fertilizer are listed in Table 10.3. The most widely used zinc fertilizer is zinc sulphate which is produced in two hydrated forms: monohydrate and heptahydrate. Zinc oxide, zinc carbonate and zinc phosphate are the other sources of zinc. Synthetic chelates such as zinc-EDTA, zinc-NTA and zinc-HEDTA are being used extensively.

Methods of Manufacture

Zinc sulphate (white vitriol) is commercially produced by treating zinc ash with sulphuric acid. Zinc ash is a by-product of galvanizing industry.

$$ZnO + H_2SO_4 = ZnSO_4 + H_2O$$

Zinc sulphate is also produced by dissolving the zinc metal scrap in hot sulphuric acid in a lead-lined vessel.

$$Zn + H_2SO_4 = ZnSO_4 + H_2$$

For production of one tonne zinc sulphate, about 250 kg zinc metal and 365 kg sulphuric acid are required.

Zinc chelates are made by complexing the metal with organic chelating agents such as EDTA, NTA and HEDTA.

Table 10.3 Common zinc fertilizers

Fertilizer material	*Chemical formula*	*Per cent zinc*	*Solubility*
Zinc sulphate heptahydrate	$ZnSO_4.7H_2O$	22	Soluble
Zinc sulphate monohydrate	$ZnSO_4.H_2O$	33	Soluble
Zinc oxide	ZnO	60–78	Insoluble
Zinc carbonate	$ZnCO_3$	52	Insoluble
Zinc phosphate	$Zn_3(PO_4)_2$	51	Insoluble
Zinc chelates	$Na_2ZnEDTA$	6–14	Soluble
	NaZnNTA	13	Soluble
	NaZnHEDTA	9	Soluble
Natural organic materials		5–10	Insoluble

Product Characteristics

Zinc sulphate is a white finely divided crystal having a specific gravity of 2.0. Zinc oxide is a white or yellowish powder which is insoluble in water. It is usually least expensive and contains more zinc than other materials. It can be added to ammonium phosphates and superphosphates. Zinc oxide is a good source of zinc in acid soils, whereas zinc sulphate monohydrate is effective as source of zinc for plants in all kinds of soils. Only zinc chelates and to a limited extent zinc sulphate are soluble enough to be used in clear liquid fertilizers. Zinc chelates are successfully added to clear liquid fertilizers.

COPPER FERTILIZERS

The common sources of copper that are used as fertilizer are listed in Table 10.4. Copper sulphate, known as blue vitriol or blue stone in pure form, is the most commonly used copper fertilizer. This is produced in two hydrated forms: monohydrate and pentahydrate. Cupric oxide (CuO), copper nitrate and copper ammonium phosphate are the other sources of copper. Synthetic chelates such as copper-EDTA and copper-HEDTA are also being used extensively.

Table 10.4 Common copper fertilizers

Fertilizer material	*Chemical formula*	*Per cent copper*	*Solubility*
Copper sulphate	$CuSO_4.5H_2O$	25	Soluble
Copper sulphate monohydrate	$CuSO_4.H_2O$	35	Soluble
Copper oxide	CuO	50–75	Insoluble
Copper nitrate	$Cu(NO_3)_2.3H_2O$	78	Soluble
Copper acetate	$Cu(C_2H_3O_2)_2.H_2O$	32	Soluble
Copper ammonium phosphate	$CuSO_4.(NH_4)PO_4.H_2O$	32	Slightly soluble
Copper chelates	$Na_2CuEDTA$	13	Soluble
	NaCuHEDTA	9	Soluble
Natural organic materials		5–7	Soluble

Methods of Manufacture

Copper sulphate (blue vitriol) is commercially produced by dissolving the metal scrap in hot sulphuric acid in a lead-lined vessel in which sufficient air is introduced to prevent loss of sulphur dioxide.

$$Cu + H_2SO_4 = CuSO_4 + H_2$$

For production of one tonne copper sulphate, about 270 kg copper metal and 420 kg sulphuric acid are required.

Cooper oxide (cupric oxide) is produced by heating copper in air or by ignition of the sulphate, nitrate, carbonate or hydroxide of the metal.

Copper chelates are made by complexing the metal with organic chelating agents such as EDTA and HEDTA.

Product Characteristics

Copper sulphate in pure form is an asymmetric crystal of light blue colour and is soluble in water. Cupric oxide has a dark brown colour and is insoluble in water.

Copper ammonium phosphate is slightly soluble in water but can be suspended and sprayed on the plants.

Copper chelates are soluble in water and can be used as foliar spray but they are more expensive than copper sulphate for foliar application as well as soil application.

MANGANESE FERTILIZERS

The common sources of manganese that are used as fertilizer are listed in Table 10.5. Manganese sulphate is the most commonly used manganese fertilizer. Manganese oxide (MnO), and manganese chloride are the other sources of manganese. Synthetic chelates such as manganese-EDTA are also used.

Table 10.5 Common manganese fertilizers

Fertilizer material	*Chemical formula*	*Per cent manganese*	*Solubility*
Manganese sulphate	$MnSO_4.4H_2O$	24–28	Soluble
Manganese oxide	MnO	41–68	Insoluble
Manganese chloride	$MnCl_2$	17	Soluble
Manganese chelates	MnEDTA	5–12	Soluble
Natural organic materials		5–9	Soluble

Methods of Manufacture

Manganese sulphate is produced by the action of sulphuric acid on pyrolusite (MnO_2).

$$2MnO_2 + 2H_2SO_4 = MnSO_4 + H_2O + O_2$$

For production of one tonne manganese sulphate, about 410 kg pyrolusite and 900 kg sulphuric acid are required.

Manganese oxide (MnO) is a by-product of the manufacture of ferromanganese metal.

Manganese chloride can be prepared by treating manganese carbonate with hydrochloric acid:

$$MnCO_3 + 2HCl \rightarrow MnCl_2 + H_2O + CO_2$$

It also forms when pyrolusite (manganese dioxide) is heated with concentrated hydrochloric acid. This reaction was once used for the manufacture of chlorine.

$$MnO_2 + 4HCl \rightarrow MnCl_2 + 2H_2O + Cl_2$$

Product Characteristics

Manganese sulphate is a purple, crystalline material which is available in several forms. Manganese oxide is a dark purple or almost black material suitable for incorporating into fertilizer.

Manganese chloride is a polymeric solid which dissolves in water to give mildly acidic solutions with a pH of around 4.

Manganese chelate absorbed on vermiculite has greenish brown colour and a volume : weight ratio about the same as a sandy loam soil.

BORON FERTILIZERS

The common sources of boron that are used as fertilizer are listed in Table 10.6. Borax (sodium tetraborate) is the most widely used boron fertilizer. Calcium borate or colemanite is another source of boron. Solubor is highly concentrated source of boron. Boric acid is readily available as pharmaceutical that can be used as boron source for plants.

Borax and boric acid are widely used in a variety of industries such as glass and ceramics, refractories, food preservation, detergents and soap, electroplating, fluxes, enamels, cosmetics, pharmaceuticals and textiles. The glass and porcelain industries are the major consumers of borax and boric acid, consuming over 50% of the total production.

Table 10.6 Common boron fertilizers

Fertilizer material	*Chemical formula*	*Per cent boron*	*Solubility*
Borax	$Na_2B_4O_7.10H_2O$	11	Soluble
Boric acid	H_3BO_3	17	Soluble
Colemanite	$Ca_2B_6O_{11}.5H_2O$	10–16	Insoluble
Sodium pentaborate	$Na_2B_{10}O_{16}.10H_2O$	18	Insoluble
Sodium tetraborate	$Na_2B_4O_7.5H_2O$	14–15	Insoluble
Solubor	$Na_2\ B_4O_7.5H_2O + Na_2B_4O_7.10H_2O$	20–21	Soluble

Methods of Manufacture

Borax is a mineral form of boron which is mined in USA, Turkey, Italy, Tibet, Chile, Argentina, Bolivia and Russia. The world's largest known reserves of borates are located in USA, supplying nearly 90-95% of the world demand.

Borax mineral contains borates of sodium (borax), calcium (colemanite) and magnesium (boracite). Borax is commercially produced by any of the following three processes:

(i) By boiling colemanite with sodium carbonate solution and crystallizing borax ($Na_2B_4O_7$) from the filtrate.

$$\underset{\text{Colemanite}}{Ca_2B_6O_{11}} + Na_2CO_3 = \underset{\text{Borax}}{N_2B_4O_7} + \underset{\text{Sodium metaborate}}{NaBO_2} + 2CaCO_3$$

The sodium metaborate ($NaBO_2$) present in the mother liquor is converted to borax by passing carbon dioxide.

$$4NaBO_2 + CO_2 = Na_2B_4O_7 + Na_2CO_3$$

(ii) By crystallizing kernite ($Na_2B_4O_7.4H_2O$): Refined borax is produced by dissolving the crude material in hot water under pressure. The insoluble impurities are filtered off and the liquor is crystallized to give borax.

(iii) Recovery from the brine containing appreciable amount of borates, for example, Searle's Lake brine in California contains 1.5% borax ($Na_2B_4O_7$).

Boric acid is prepared by the action of sulphuric acid on colemanite or concentrated solution of borax.

$$Ca_2B_6O_{11} + 2H_2SO_4 + 5H_2O = H_4B_6O_{11} + 2CaSO_4$$
$$H_4B_6O_{11} + 7H_2O = 6H_3BO_3$$
$$Na_2B_4O_7 + H_2SO_4 + 5H_2O = 4H_3BO_3 + Na_2SO_4$$

It is also obtained from the volcanic hot springs of Tuscany, Italy. The hot liquid is evaporated by the heat of steam jets and boric acid crystallizes out in thin white plates. Solubor is made by refining borax with colemanite.

Product Characteristics

Borax (sodium tetraborate decahydrate) is a white or light gray salt finely ground for fertilizer use. It is soluble in water. Solubor is also a finely ground white material which is soluble in water. Colemanite is white crystalline salt that is slightly soluble in water but soluble in weak acids.

Boric acid is a white powder usually offered as a pharmaceutical product and it is readily soluble in water.

CHLORINE FERTILIZERS

Chlorine is required by plants as chloride ion which is usually supplied in sufficient amount by soil, irrigation or rainwater or through incidental addition. Some plants require more chlorine than others, for example, beets and tomatoes and they need supplemental addition of chloride. Chloride fertilizers are not commonly applied to crops.

Several soluble sources of chlorine are available (Table 10.7). Ammonium chloride and potassium chloride (Muriate of potash) are popular fertilizers supplying nitrogen and potassium, respectively. Calcium chloride and sodium chloride (common salt) are the other easily available sources of chloride.

Table 10.7 Common chlorine fertilizers

Fertilizer material	*Chemical formula*	*Per cent chlorine*	*Solubility*
Ammonium chloride	NH_4Cl	66	Soluble
Calcium chloride	$CaCl_2$	65	Insoluble
Potassium chloride	KCl	74	Soluble
Sodium chloride	NaCl	47	Soluble

MOLYBDENUM FERTILIZERS

The common sources of molybdenum that are used as fertilizer are listed in Table 10.8. Ammonium molybdate and sodium molybdate are commonly used molybdenum fertilizers. Molybdenum trioxide is another rich source. Frits containing molybdenum are also available for use as molybdenum source for plants.

Table 10.8 Common molybdenum fertilizers

Fertilizer material	*Chemical formula*	*Per cent molybdenum*	*Solubility*
Ammonium molybdate	$(NH_4)_6Mo_7O_{24}.2H_2O$	54	Soluble
Sodium molybdate	$Na_2MoO_4.2H_2O$	39	Soluble
Molybdenum trioxide	MoO_3	66	Insoluble
Molybdenum frits	Fritted glass	1–30	Insoluble

Methods of Manufacture

Molybdenum trioxide is prepared by roasting the sulphide ore of molybdenum (MoS_2) in air or by action of nitric acid on molybdenum metal. Ammonium molybdate is obtained by the treatment of molybdenum trioxide with ammonium hydroxide followed by evaporation and crystallization. Sodium molybdate is prepared in a similar manner.

Product Characteristics

Ammonium molybdate and sodium molybdate are white powders which are soluble in water.

Molybdenum trioxide is insoluble in water. Molybdenum frits are glassy materials, which release molybdenum very slowly.

Selected References

Havlin, J.L., Beaton, J.D., Tisdale, S.L. and Nelson, W.L. (2005). *Soil Fertility and Fertilizers*, 7th ed. Pearson Education, Upper Saddle River, N.J.

Mordtvet, J.J., Murphy, L.S. and Follett, R.F.H. (1999). *Fertilizer Technology and Application*. Meister Publishing Company, Euclid Ave., Willoughby, OH.

UNIDO/IFDC, Ed. (1998). *Fertilizer Manual*, 3rd ed., Kluwer Academic Publishers, Dordrecht, the Netherlands.

CHAPTER

11

Fertilizer Legislation and Quality Control

Fertilizer Legislation

Fertilizers are used by farmers who are commonly not capable of evaluating the standards and quality of the fertilizers they buy from the retailer. If the fertilizer is not of the desired standard or it does not contain the nutrient to the extent it should, its use is not likely to give the expected increase in crop yield. This will result in yield loss and economic loss to the farmers. In order to protect the interest of farmers, regulations have been established in many countries to govern marketing of commercial fertilizers at retail level. These regulations are tailored to local conditions and thus vary from one country to the other. However, the primary purpose remains the same, that is, to ensure uniformity of the quality and to provide a choice to the farmer to select the most economical fertilizer product from the market.

Enforcement of regulations adds to the cost of fertilizers. In order to guarantee a given nutrient percentage, the manufacturer must provide some surplus amount of nutrient, to take care of sampling and analytical errors.

In India sale, distribution and quality of fertilizers is governed by the Fertilizer Control Order 1957 issued by the Government of India under the Essential Commodities Act. Fertilizer is thus an essential commodity. The new Fertilizer Control Order (FCO) was issued in 1985 replacing the earlier order. The order has been amended many times as per need of the fertilizer situation.

The FCO performs five functions:

1. It defines the 'fertilizer' and providers a list of materials which can be labelled and sold as fertilizers.
2. It makes it compulsory for anyone interested in producing, importing or selling fertilizer to get an authorization from the state/central govt. to do so.
3. It specifies the information to be provided on the fertilizer bag and sets up the norms for fertilizer trading, registration/authorization of dealers, record-keeping and duties of fertilizer inspectors and dealers.

4. It specifies the physical and chemical properties of each listed fertilizer in quantitative terms for maintaining quality.
5. It provides the procedure for collection of fertilizer sample and analysis for quality control.

Fertilizer Specification

Specifications are the requirements with which a fertilizer should conform as per local legislation or as agreed upon between buyer and seller.

Normally, the standard specifications are printed on the fertilizer bag as per requirements of Government regulations. This consists of fertilizer grade, net weight, name of manufacturer or importer and sometimes additional information about the product. This is basic information about the product which does not give details of composition or quality.

A complete fertilizer specification should include:

1. Nutrient content: major nutrients (total N, P_2O_5 and K_2O), sulphur and micronutrients guaranteed by the manufacturer (% minimum).
2. Moisture content (% maximum at the time of filling the bag)
3. Particle size distribution (% of small and large size particles)
4. Physical condition (granular or powder)
5. Solubility and availability (water solubility of phosphate, citrate solubility of insoluble phosphate fertilizers)
6. Conditioner (name and content)
7. Specific limitations: presence of toxic substances like biuret in urea
8. Packaging details, if any

Fertilizers in FCO

Since promulgation of FCO in 1957 by the Government of India, the number of fertilizers included in the list has grown to over 70. Chemical and physical specifications are laid out for each fertilizer. The list includes only straight and compound NPK fertilizers and sources of sulphur and micronutrients. The specifications of multi-micronutrient mixtures are decided by the state governments. There are 80 approved multi-micronutrient mixtures approved by various states in India. Specifications of important fertilizers are given in Table 11.1 and the tolerance limits in plant nutrients and physical parameters for various fertilizers are given in Table 11.2.

Table 11.1 Specifications of important fertilizers (as per FCO 1985)

1.	**Ammonium sulphate**	
	(i) Moisture, per cent by weight, maximum	1.0
	(ii) Ammoniacal nitrogen, per cent by weight, minimum	20.6
	(iii) Free acidity (as H_2SO_4), per cent by weight, maximum (0.04 for material obtained from by-product ammonia and by-product gypsum)	0.025
	(iv) Arsenic (as As_2O_3), per cent by weight, maximum	0.01
	(v) Sulphur (as S), per cent by weight, minimum	23.0

2.	**Urea**	
	(i) Moisture, per cent by weight, maximum	1.0
	(ii) Total nitrogen, per cent by weight (on dry basis), minimum	46.0
	(iii) Biuret, per cent by weight, maximum	1.5
	(iv) Particle size—Not less than 90 per cent of the material shall pass through 2.8 mm IS sieve and not less than 80 per cent by weight shall be retained on 1 mm IS sieve	
3.	**Urea (Granular)**	
	(i) Moisture, per cent by weight, maximum	1.0
	(ii) Total nitrogen, per cent by weight (on dry basis), minimum	46.0
	(iii) Biuret, per cent by weight, maximum	1.5
	(iv) Particle size—Not less than 90 per cent of the material shall pass through 4 mm IS sieve and be retained on 2 mm IS sieve. Not more than 5 per cent shall be below 2 mm IS sieve.	
4.	**Single superphosphate (16% P_2O_5 Powder)**	
	(i) Moisture, per cent by weight, maximum	12.0
	(ii) Free phosphoric acid (as P_2O_5), per cent by weight, maximum	4.0
	(iii) Water-soluble phosphates (as P_2O_5), per cent by weight, minimum	16.0
	(iv) Sulphur (as S), per cent by weight, minimum	11.0
5.	**Single superphosphate (16% P_2O_5 Granulated)**	
	(i) Moisture, per cent by weight, maximum	5.0
	(ii) Free phosphoric acid (as P_2O_5), per cent by weight, maximum	4.0
	(iii) Water-soluble phosphates (as P_2O_5), per cent by weight, minimum	16.0
	(iv) Particle size—Not less than 90 per cent of the material shall pass through 4 mm IS sieve and shall be retained on 1 mm IS sieve. Not more than 5 per cent shall pass through 1 mm IS sieve.	
	(v) Sulphur (as S), per cent by weight, minimum	11.0
6.	**Potassium chloride (Muriate of Potash)**	
	(i) Moisture, per cent by weight, maximum	0.5
	(ii) Water-soluble potash content (as K_2O), per cent by weight, minimum	60.0
	(iii) Sodium as NaCl, per cent by weight (on dry basis), maximum	3.5
	(iv) Particle size—Minimum 65 per cent of the material shall pass through 1.7 mm IS sieve and be retained on 0.25 mm IS sieve.	
7.	**Potassium sulphate**	
	(i) Moisture, per cent by weight, maximum	0.5
	(ii) Potash content (as K_2O), per cent by weight, minimum	50.0
	(iii) Total chloride (as Cl), per cent by weight (on dry basis), maximum	2.5
	(iv) Sodium as NaCl, per cent by weight (on dry basis), maximum	2.0
	(v) Sulphur (as S), per cent by weight, minimum	17.5

8.	**Diammonium phosphate (18-46-0)**	
(i)	Moisture, per cent by weight, maximum	1.5
(ii)	Total nitrogen, per cent by weight, minimum	18.0
(iii)	Ammoniacal nitrogen form, per cent by weight, minimum	15.5
(iv)	Total nitrogen in the form of urea, per cent by weight, maximum	2.5
(v)	Neutral ammonium citrate-soluble phosphates (as P_2O_5), per cent by weight, minimum	46.0
(vi)	Water-soluble phosphates (as P_2O_5), per cent by weight, minimum	41.0
(vii)	Particle size—Not less than 90 per cent of the material shall pass through 4 mm IS sieve and be retained on 1 mm IS sieve. Not more than 5 per cent shall be below 1 mm IS sieve	
9.	**Ammonium phosphate sulphate (16-20-0)**	
(i)	Moisture, per cent by weight, maximum	1.0
(ii)	Total ammoniacal nitrogen, per cent by weight, minimum	16.0
(iii)	Neutral ammoniacal citrate-soluble phopshates (as P_2O_5), per cent by weight, minimum	20.0
(iv)	Water-soluble phosphates (as P_2O_5) , per cent by weight, minimum	19.5
(v)	Particle size—Not less than 90 per cent of the material shall pass through 4 mm IS sieve and shall be retained on 1 mm IS sieve. Not more than 5 per cent shall be below 1 mm IS sieve	
(vi)	Sulphur (as S), per cent by weight, minimum	13.0
10.	**Nitro phosphate (20-20-0)**	
(i)	Moisture, per cent by weight, maximum	1.5
(ii)	Total nitrogen, per cent by weight, minimum	20.0
(iii)	Nitrogen in ammoniacal form, per cent by weight, minimum	10.0
(iv)	Nitrogen in nitrate form, per cent by weight, maximum	10.0
(v)	Neutral ammonium citrate-soluble phosphates (as P_2O_5), per cent by weight, minimum	20.0
(vi)	Water-soluble phosphates (as P_2O_5), per cent by weight, minimum	12.0
(vii)	Calcium nitrate, per cent by weight, max.	1.0
(viii)	Particle size—Not less than 90 per cent of the material shall pass through 4 mm IS sieve and be retained on 1 mm IS sieve. Not more than 5 per cent shall be below 1 mm IS sieve	
11.	**Urea-Ammonium phosphate (28-28-0)**	
(i)	Moisture, per cent by weight, maximum	1.5
(ii)	Total nitrogen, per cent by weight, minimum	28.0
(iii)	Ammoniacal nitrogen, per cent by weight, minimum	9.0
(iv)	Neutral ammonium citrate-soluble phosphates (as P_2O_5), per cent by weight, minimum	28.0
(v)	Water-soluble phosphates (as P_2O_5), per cent by weight, minimum	25.2
(vi)	Particle size—Not less than 90 per cent of the material shall pass through 4 mm IS sieve and shall be retained on 1 mm IS sieve. Not more than 5 per cent shall be below 1 mm IS sieve	

12.	**Mono ammonium phosphate (11-52-0)**	
	(i) Moisture, per cent by weight, maximum	1.0
	(ii) Total nitrogen all in ammoniacal form, per cent by weight, minimum	11.0
	(iii) Neutral ammonium citrate-soluble phosphates (as P_2O_5), per cent by weight, minimum	52.0
	(iv) Water-soluble phosphates (as P_2O_5), per cent by weight, minimum	44.2
	(v) Particle size—Not less than 90 per cent of the material shall pass through 4 mm IS sieve and shall be retained on 1 mm IS sieve. Not more than 5 per cent shall be below 1 mm IS sieve	
13	**NPK (15-15-15)**	
	(i) Moisture, per cent by weight, maximum	1.5
	(ii) Total nitrogen, minimum	15.0
	(iii) Ammoniacal nitrogen, per cent by weight, minimum	7.5
	(iv) Nitrate nitrogen, per cent by weight, maximum	7.5
	(v) Neutral ammonium citrate-soluble phosphates (as P_2O_5), per cent by weight, minimum	15.0
	(vi) Water-soluble phosphates (as P_2O_5), per cent by weight, minimum	4.0
	(vii) Water-soluble potash (as K_2O), per cent by weight, minimum	15.0
	(viii) Particle size—Not less than 90 per cent of the material shall pass through 4 mm IS sieve and be retained on 1 mm IS sieve.	
	(ix) Calcium nitrate, per cent by weight, maximum	1.0
14.	**NPK (10-26-26)**	
	(i) Moisture, per cent by weight, maximum	1.0
	(ii) Total nitrogen, per cent by weight, minimum	10.0
	(iii) Ammoniacal nitrogen, per cent by weight, minimum	7.0
	(iv) Nitrogen in the form of urea, per cent by weight, maximum	3.0
	(v) Neutral ammonium citrate-soluble phosphates (as P_2O_5), per cent by weight, minimum	26.0
	(vi) Water-soluble potash (as K_2O), per cent by weight, minimum	26.0
	(vii) Water-soluble phosphates (as P_2O_5), per cent by weight, minimum	22.1
	(viii) Particle size—Particle size of the material will be such that 90 per cent of the material will be between 1 mm and 4 mm IS sieve and not more than 5 per cent will be below 1 mm size.	
15.	**NPK (12-32-16)**	
	(i) Moisture, per cent by weight, maximum	1.0
	(ii) Total nitrogen, per cent by weight, minimum	12.0
	(iii) Ammoniacal nitrogen, per cent by weight, minimum	9.0
	(iv) Nitrogen in the form of urea, per cent by weight, maximum	3.0
	(v) Neutral ammonium citrate-soluble phosphates (as P_2O_5), per cent by weight, minimum	32.0
	(vi) Water-soluble phosphates (as P_2O_5), per cent by weight, minimum	27.2
	(vii) Water-soluble potash (as K_2O), per cent by weight, minimum	16.0
	(viii) Particle size—Particle size of the material will be such that 90 per cent of the material will be between 1 mm and 4 mm IS sieve and not more than 5 per cent will be below 1 mm size.	

16.	**Zinc sulphate-Heptahydrate ($ZnSO_4.7H_2O$)**	
	(i) Matter insoluble in water, per cent by weight, maximum	1.0
	(ii) Zinc (as Zn), per cent by weight, minimum	21.0
	(iii) Lead (as Pb), per cent by weight, maximum	0.003
	(iv) Copper (as Cu), per cent by weight, maximum	0.1
	(v) Magnesium (as Mg), per cent by weight, maximum	0.5
	(vi) pH, not less than	4.0
	(vii) Sulphur (as S), per cent by weight, minimum	10.0
	(viii) Cadmium (as Cd), per cent by weight, maximum	0.0025
	(ix) Arsenic (as As), per cent by weight, maximum	0.01
17.	**Borax (Sodium tetraborate) ($Na_2B_4O_7.10H_2O$) for soil application**	
	(i) Content of boron (as B), per cent by weight, minimum	10.5
	(ii) Matter insoluble in water, per cent by weight, maximum	1.0
	(iii) pH	9.0-9.5
	(iv) Lead (as Pb), per cent by weight, maximum	0.003
PROVISIONAL FERTILIZERS UNDER CL. 20 A		
1.	**NPK Complex fertilizer fortified with Zinc (12:32:16:0.5)**	
	(i) Moisture, per cent by weight, maximum	1.5
	(ii) Total nitrogen, per cent by weight, minimum	12.0
	(iii) Ammoniacal nitrogen, per cent by weight, minimum	9.0
	(iv) Urea nitrogen (as N), per cent by weight, maximum	3.0
	(v) Neutral ammonium citrate-soluble phosphate (as P_2O_5), per cent by weight, minimum	32.0
	(vi) Water-soluble phosphate (as P_2O_5) as per cent by weight, minimum	25.0
	(vii) Water-soluble potash (as K_2O), per cent by weight, minimum	16.0
	(viii) Zinc (as Zn), per cent by weight, minimum	0.5
	(ix) Particle size—Not less than 90 per cent of the material shall be between 1 mm and 4 mm IS sieve and not more than 5 per cent shall be below 1 mm IS sieve	
2.	**NPK Complex fertilizer fortified with Zinc (10:26:26:0.5)**	
	(i) Moisture, per cent by weight, maximum	1.5
	(ii) Total nitrogen, per cent by weight, minimum	10.0
	(iii) Ammoniacal nitrogen, per cent by weight, minimum	7.0
	(iv) Urea nitrogen (as N), per cent by weight, maximum	3.0
	(v) Neutral ammonium citrate-soluble phosphate (as P_2O_5), per cent by weight, minimum	26.0
	(vi) Water-soluble phosphate (as P_2O_5), per cent by weight, minimum	20.0
	(vii) Water-soluble potash (as K_2O), per cent by weight, minimum	26.0
	(viii) Zinc (as Zn), per cent by weight, minimum	0.5
	(ix) Particle size—Not less than 90 per cent of the material shall be between 1 mm and 4 mm IS sieve and not more than 5 per cent shall be below 1 mm IS sieve	

3.	**Neem-coated urea**	
	(i) Moisture, per cent by weight, maximum	1.0
	(ii) Total nitrogen, per cent by weight, minimum	46.0
	(iii) Biuret, per cent by weight, maximum	1.5
	(iv) Benzene-soluble content, per cent by weight, minimum	0.035
	(v) Particle size—Not less than 90 per cent of the material shall pass through 2.8 mm IS sieve and not less than 80 per cent by weight shall be retained on 1 mm IS sieve	
4.	**NPK Complex fertilizer fortified with Boron (12:32:16:0.3)**	
	(i) Moisture, per cent by weight, maximum	1.0
	(ii) Total nitrogen, per cent by weight, minimum	12.0
	(iii) Ammoniacal nitrogen, per cent by weight, minimum	9.0
	(iv) Urea nitrogen (as N), per cent by weight, maximum	3.0
	(v) Neutral ammonium citrate-soluble phosphate (as P_2O_5), per cent by weight, minimum	32.0
	(vi) Water-soluble phosphate (as P_2O_5), per cent by weight, minimum	27.2
	(vii) Water-soluble potash (as K_2O), per cent by weight, minimum	16.0
	(viii) Boron (as B), per cent by weight	0.3
	(ix) Particle size—Not less than 90 per cent of the material shall be between 1 mm and 4 mm IS sieve and not more than 5 per cent shall be below 1 mm IS sieve	
5.	**NPK Complex fertilizer fortified with Boron (10:26:26:0.3)**	
	(i) Moisture, per cent by weight, maximum	1.0
	(ii) Total nitrogen, per cent by weight, minimum	10.0
	(iii) Ammoniacal nitrogen, per cent by weight, minimum	7.0
	(iv) Urea nitrogen (as N), per cent by weight, maximum	3.0
	(v) Neutral ammonium citrate-soluble phosphate (as P_2O_5), per cent by weight, minimum	26.0
	(vi) Water-soluble phosphate (as P_2O_5) as per cent by weight, minimum	22.1
	(vii) Water-soluble potash (as K_2O), per cent by weight, minimum	26.0
	(viii) Boron (as B), per cent by weight	0.3
	(ix) Particle size—Not less than 90 per cent of the material shall be between 1 mm and 4 mm IS sieve and not more than 5 per cent shall be below 1 mm IS sieve	
6.	**Diammonium phosphate fortified with Boron (18:46:0:0.3)**	
	(i) Moisture, per cent by weight, maximum	1.5
	(ii) Total nitrogen, per cent by weight, minimum	18.0
	(iii) Ammoniacal nitrogen, per cent by weight, minimum	15.5
	(iv) Total nitrogen in the form of urea, per cent by weight, maximum	2.5
	(v) Neutral ammonium citrate-soluble phosphate (as P_2O_5), per cent by weight, minimum	46.0
	(vi) Water-soluble phosphate (as P_2O_5), per cent by weight, minimum	41.0
	(vii) Boron (as B), per cent by weight	0.3
	(viii) Particle size—Not less than 90 per cent of the material shall pass through 4 mm IS sieve and be retained on 1 mm IS sieve. Not more than 5 per cent shall be below 1 mm size.	

7.	**Bentonite-Sulphur**	
	(i) Moisture, per cent by weight, maximum (ii) Total sulphur (as S), per cent by weight, minimum (iii) Particle size—Not less than 90 per cent of the material shall pass through 4 mm sieve and be retained on 1 mm IS sieve and not more than 5% shall be below 1 mm IS sieve. Note: The fertilizer may contain about 10% bentonite as filler.	0.5 90.0

Table 11.2 Tolerance limit in plant nutrients and physical parameters for various fertilizers

S.No.	*Fertilizer*	*Tolerance limit*	
1. (a)	For fertilizers with definite compounds like ammonium sulphate, urea, ammonium chloride, muriate of potash, superphosphate containing more than 20 per cent plant nutrients	0.2 unit	
(b)	Containing less than 20 per cent plant nutrients	0.1 unit	
2.	Calcium ammonium nitrate	0.3 unit	
3.	Diammonium phosphate	0.5 units each for N and P content	
4.	Nitrophosphate, ammonium sulphate nitrate, urea-ammonium phosphate, ammonium phosphate sulphate, compound/complex fertilizers, granular and physical mixtures of fertilizers	Total tolerance limit shall be subject to maximum of 2% of all combined nutrients but for individual nutrients, it will vary as under	
		Nutrient level (%)	*Tolerance level (unit)*
		15 or less 16 to 20 21 or more	0.5 0.6 0.7
5.	Borax, chelated zinc-EDTA and chelated iron-EDTA, magnesium sulphate	0.1 unit	
6.	Copper sulphate, zinc sulphate, manganese sulphate and ferrous supphate	0.2 unit	
7.	Ammonium molybdate	0.5 unit	
8.	Mixture of micronutrient fertilizers	*Combined nutrient level (%)(sum of all micronutrients)*	*Tolerance level (unit)*
		10 or less 11 to 20 21 or more	0.1 0.2 0.5
9.	Particle size	$\pm$3 unit	
10.	Moisture	0.3 unit	

Quality Control

Quality control is a procedure or set of procedures intended to ensure that a manufactured product adheres to a defined set of quality criteria or meets the requirements of the regulatory agency or customer.

Fertilizer quality standards are enforced by the government and the compliance of the quality standards is ensured by the manufacturer. To ensure the quality of the fertilizer that is produced, manufacturers monitor the product at each stage of production. The raw materials and the finished products are all subjected to a series of physical and chemical tests to ensure that they meet the standard specifications. Since fertilizer production is regulated by the government, composition analysis tests are run on samples to determine total nitrogen, phosphate and potash content, and other elements affecting the chemical composition. Several other tests are also performed, depending on the specific nature of the fertilizer product.

Fertilizer Quality Control by Manufacturers

Fertilizer manufacturers or importers have the responsibility for ensuring the good quality of fertilizers from factory gate to retail sale point. This is a great responsibility to be discharged under supervision of the government. Frequent analysis of raw materials, process materials and end product is required for ensuring quality inputs, adjustment of the process in time and ensuring product on grade. The control measures, which go in the production of products having high quality, are summarized below:

1. **Raw material quality**
   Control over the nutrients in the end product has its beginning from the quality assurance of the raw materials. Hence, all raw materials are analyzed to conform to stipulated specifications before acceptance and storage. The particle size of the material is an important factor in determining its behaviour during handling and granulation. The analytical control is also exercised on the quality of filler, other inputs and utilities in case of production of NPK mixed fertilizers.
2. **Process control**
   Sampling and analysis of the process materials is important for the production of high quality fertilizers. The analytical methods devised for process control should be rapid, precise, and reliable. Various process parameters are checked frequently for adjustments, whenever necessary.
3. **Product quality control**
   In order to ensure that the fertilizer product meets the standard specifications before it is dispatched to the distributors and retailers, analysis of each batch of the product is carried out and only the lots passing the test are allowed for dispatch to store or delivery system.
4. **Quality control during storage**
   The products from the fertilizer plant are stored in bulk storage or bagged directly. The bulk is reclaimed with scraper loaders moving continuously along the length of the bins to get homogenous and blended composite material for final dispatch. The blending operation

helps in removing non-uniformity in the composition of the product received at different manufacturing periods from the plant.

The bagged product is analyzed for all the nutrients before dispatch. The process further improves the nutrient control of the product for compliance of FCO tolerance limit in plant nutrients.

5. **Field level quality control**

   Fertilizer manufacturers exercise rigid surveillance measures through their marketing unit to ensure that the quality of the product does not deteriorate during transportation, storage and handling at field level. Samples of the product are drawn on random basis by the field officials and sent back to plant laboratory for nutrient check. These samples are collected from the entire area of marketing operations all over the country. Thus, a feedback is available to the fertilizer plant about the status of quality of the product at the retail sale point, and appropriate actions are taken to maintain the desired level of nutrients in the product upto retail sale point.

Fertilizer Quality Control Enforcement by the Government

The fertilizer samples are drawn from the sale points by the government officials and sent to the nearest fertilizer quality control laboratory for analysis and report whether the fertilizer quality especially the nutrient content meets the standard specification prescribed in the Fertilizer Control Order, 1985. Presently, there are 67 fertilizer quality control laboratories in India (4 Central Government laboratories and 63 State Government laboratories) with a total annual analyzing capacity of 125 thousand samples. The analytical capacity and the number of samples analyzed and found non-standard during the last 5 years are given in Table 11.3.

Table 11.3 Analytical capacity of fertilizer quality control laboratories and the number of fertilizer samples analyzed

Year	*Number of laboratories*	*Annual analytical capacity*	*Number of samples analyzed*	*% of non-standard samples*
2000-01	66	120315	103436	5.3
2001-02	66	120415	104188	5.7
2002-03	67	125480	109504	5.4
2003-04	67	124778	104647	5.5
2004-05	67	124730	108859	6.0

The percentage of non-standard fertilizer samples on all India basis is around 5-6 per cent. However, there is a large variation from state to state—from as high as 18 per cent in Madhya Pradesh to 1-3 per cent in Punjab and Gujarat.

Selected References

FAI (2001). *Handbook on Fertilizer Technology*. 6th ed., Fertiliser Association of India, New Delhi, pp. 273.

FAI (2005). *Fertiliser Control Order* (1985). Fertiliser Association of India, New Delhi.

CHAPTER

12

Fertilizer Marketing

Marketing is defined as "the total of activities involved in the transfer of goods from the producer or seller to the consumer or buyer, including advertising, shipping, storing, and selling."

The marketing can be categorized as a branch of business as well as a social science. People buy goods (thus becoming the buyer/consumer) from a vendor (or producer/seller), creating a transaction. In the past, marketing involved travelling salesmen, while in modern times, marketing is more likely to involve print media, television, the Internet, and other forms of media made available by the information and communication technology.

FERTILIZER MARKETING

As fertilizer marketing deals with a key agricultural input which is needed by rural people engaged in agriculture, the strategy has to be different than for common consumer goods. Fertilizer marketing operates in an environment which is influenced by many external factors like distributors, retailers, transporters, govt. extension personnel, competitors and customers. Each one has its perception of the situation which in turn influences its decision-making. The marketing network plays an important role in ensuring fertilizer availability to the farmers spread over various parts of the country. Unlike consumer goods, fertilizer distribution system is governed by certain unique features such as:

1. Fertilizer production is continuous and demand is not only seasonal but short-lived.
2. Fertilizer is distributed through a huge network of retail outlets (over 292 thousand in India) spread in various parts of the country.
3. Government controls distribution and movement of some fertilizers.
4. Fertilizer production is concentrated in limited area.
5. There are inadequate storage facilities in rural area where fertilizer is used.
6. There are a large number of users with different consumer profile base.

Marketing Channels

The basic fertilizer distribution channel is depicted in Fig. 12.1. The most common channel is from the supplier (manufacturer or importer) to the wholesaler and on to the retailer. However, a manufacturer or importer can also directly supply to a retailer. A retailer may be a private party or cooperative society. The end user of fertilizers is the farmer who gets the desired material from the retailer. Some fertilizer firms sell their product directly to a limited extent to farmers through service centres operated by the company itself.

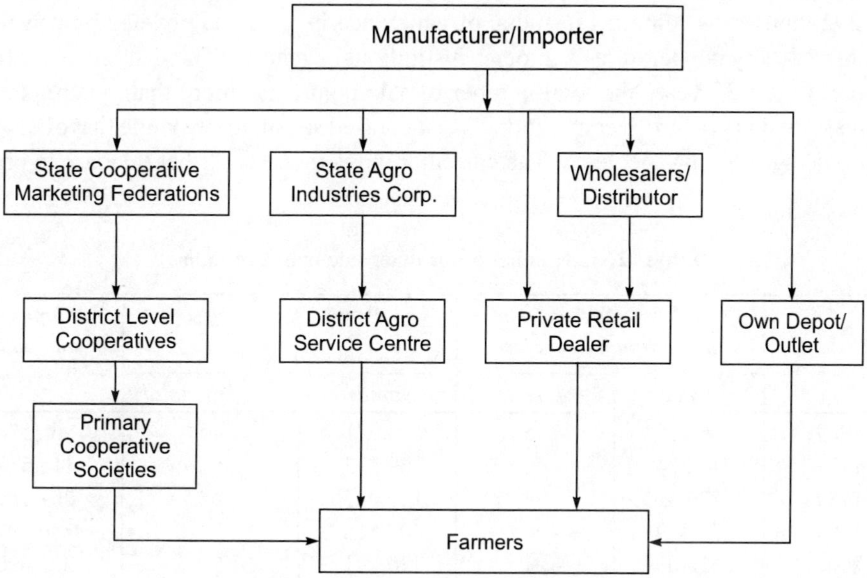

Fig. 12.1 Fertilizer marketing channels

There are several agencies functioning as intermediary in the fertilizer marketing channel. They are:

State Cooperative Marketing Federations: These are apex bodies at the state level procuring large amounts of fertilizers from the manufacturers or importers. They forward the material to district level cooperatives which distribute the fertilizers to primary cooperative societies with farmers as members. The farmer members get fertilizers from the cooperative society on credit.

State Agro Industries Corporations: These are also apex bodies at the state level procuring large amounts of fertilizers from the manufacturers or importers. They distribute the material to district level agro service centres which sell the products to farmers on cash payment or on credit through banks.

Wholesalers/Distributors: These are private agencies which play most important role in fertilizer marketing as they handle larger quantities of fertilizers than any other intermediary agency. They provide efficient service in fertilizer marketing by catering to the requirements of the network of private retailers spread over entire country.

Manufacturer's Depots/Outlets: Some manufacturers maintain their own depots or outlets for sale of fertilizers to farmers. These outlets may be in the form of farmers' service centres operated by the fertilizer company or a franchisee.

Retail Network

The retail network provides the fertilizer sale points from where the farmers get the needed fertilizers. The sale points are either managed by institutional agencies such as cooperatives, agro service centres and farmer's service centres or by private dealers and retailers. In 2006, there were more than 292 thousand sale points in India, of which nearly 79% were owned by private dealers and remaining 21% by cooperatives and other institutional agencies. The data given in Table 12.1 show that during last 35 years the total number of sale points has more than quadrupled and the share of private retailers in fertilizer marketing has increased substantially while that of cooperatives and institutional agencies has declined. The current trend is towards further increase in private sale points.

Table 12.1 Number of fertilizer sale points in India

As on	*Cooperatives and other institutional agencies*		*Private sale points*		*Total number of sale points*
	Number	*% of the total*	*Number*	*% of the total*	
March 1969	36,505	55	30,071	45	66,576
March 1979	46,224	40	69,293	60	115,517
March 1989	72,142	35	135,047	65	207,189
March 1999	72,579	26	207,818	74	280,397
March 2006	62,401	21	230,291	79	292,692

Source: Fertiliser Statistics 2006-07.

In India, cooperative sector is actively involved in fertilizer production and distribution. The cooperatives have a widespread network of retail outlets throughout the country. However, the cooperative movement is stronger in western India than in other parts. The small and marginal farmers are generally not members of the cooperative societies; hence, they do not get the benefits of credit facilities of the cooperative societies for purchase of fertilizers. Large and medium farmers usually take all the advantages offered by cooperative societies.

Farmer's Service Centres

Among various types of fertilizer sale points, the service centres represent a system which has an inbuilt provision for providing inputs as well as services and technical guidance at a single window. These centres also serve as informal contact point between the end user and the fertilizer producer. The farmer's service centres are operated by several fertilizer companies and the state agro industries corporations. More than 600 farmer's service centres are in operation with several names such as Farm Super Market, Farmers Shopping Centres, Krishi Seva Kendra, etc.

FORECASTING FERTILIZER DEMAND

Fertilizer is a key input for agricultural production and it is a cost-sensitive product. The estimates of fertilizer demand, if known in advance, influence the decisions concerning production, procurement, distribution and promotion. This will assure sufficient supply to the farmers and minimize the inventories, avoid running out of stock and, generally, increase sales and improve profits.

For a fertilizer manufacturer, estimates of demand and likely market share are indispensable for strategic decisions and those concerning resource allocation. Forecasts of fertilizer demand in the short term and potential consumption in the long run are needed for the determination of appropriate policies regarding food production and fertilizer use at the state and national level.

Need of Demand Forecasting

Fertilizer demand depends on many agro-economic factors. The forecasting of fertilizer demand is important for successful operation of fertilizer companies and for the formulation of appropriate policies by governments.

To arrange timely supplies of the fertilizers in thousands of villages, it is necessary to have an estimate of fertilizer demand at numerous locations at different times in both the short and medium terms. Effective demand forecasting can enable importers to take full advantage of world market price fluctuations. Required storage, transport and financial arrangements are dependent on demand. If actual fertilizer demand is less than the fertilizer produced in or imported into a country, heavy financing costs and product losses will occur as the fertilizer procured but not sold may have to be kept for six months or a year before it finds a buyer and that storage can cause some loss in quantity and quality.

The fertilizer manufacturing or marketing companies develop their action plan directly or indirectly from the demand forecast. The sales forecast advises the production unit what to produce, how much and when. For imported fertilizer, the sales forecast is the basis for the procurement schedule. With the help of a detailed sales estimate by region and month, it is possible to identify primary storage locations in such a manner as to optimize transport cost, minimize storage duration and avoid wasteful inter-store movements.

Short- and long-term forecasts serve different purposes. Short-term forecasts are concerned with the next season or year. Short-term projections are needed to arrange production, procurement of finished product, raw materials and packing materials, storage, transport and working capital so that the expected demand is met successfully. Long-term projections over four to five years are used for policy and investment decisions. Investment in new production facilities, formulation of development plans, strengthening credit institution, etc. are examples of decisions determined by long-term forecasts.

Factors Affecting Fertilizer Demand

Use of fertilizer by the farmers is like an investment. The benefit from fertilizer application should be high enough to motivate farmers to use fertilizer. Both the quantum of the benefit and its awareness by farmers are equally important to create fertilizer demand.

The farmer's purchasing power for fertilizers depends on two factors, viz. affordability and cash liquidity. The input-output price ratio, i.e. the number of kilograms of produce required to buy a kilogram of fertilizer/nutrient, is one indicator of affordability. For a given fertilizer application, if the yield increases due to better seeds, the higher is the return to the farmer and, therefore, the more affordable is the fertilizer. Prices are important means of making fertilizer application attractive.

In developing countries characterized by small holdings and subsistence farming, fertilizer demand tends to be price sensitive. When the fertilizer price goes up in relation to the produce price, consumption tends to fall in the short run until technological adjustments or other factors improve yield response to neutralize the effect of the price increase.

Adequate overall availability, good infrastructure and effective logistics management are essential to convert the forecast demand into sales. Farmers' demand cannot be met unless fertilizer is transported and stored with wholesalers and retailers in time for seasonal application.

Rainfall and its distribution are known to determine the crop growth and fertilizer consumption. Even if all other factors are favourable, failure of seasonal rains adversely affects fertilizer demand. Failure of rains increases the risk of using fertilizer and other expensive inputs, thus making the farmers reluctant to invest in fertilizers. Fertilizer demand is significantly higher in irrigated regions compared to rainfed regions.

Other factors which influence the fertilizer demand are soil characteristics and nutrient content, cropping pattern and adoption of seeds of high yielding varieties. Soil fertility status is an important consideration in prescribing fertilizer doses to crops. For example, if the soils of a given area are generally deficient in phosphorus, demand for phosphate fertilizers will be high. On the contrary, if the soils of the region are generally rich in potash, there will be little demand for potash fertilizers. Demand for nitrogen fertilizers is universal because all the soils are deficient in this nutrient and unable to meet the high requirement of non-leguminous crops. Fertilizer demand is high in areas practising intensive agriculture and in the area growing high value cash crops like sugar cane. Increase in the area under high yielding varieties of the crops triggers increase in fertilizer demand.

Forecasting Methods

There are many methods of fertilizer demand forecasting. Some are simple while others are too sophisticated. Common methods are discussed briefly here.

Time Series Analysis and Trend Extension Method

A common method of demand forecasting is the analysis of historical data to discern the trend in demand growth and extend it into the future to forecast demand. This method calls for application of statistical techniques. If data of fertilizer sales for several years are available and the trend is relatively stable, it is possible to predict the demand growth and the rate at which the growth is increasing or decreasing.

The trend extension system may be further refined to incorporate the fertilizer price index, irrigation factors and agricultural income. The overall fertilizer (N, P and K) forecast is made and individual nutrient demand is derived on the basis of the ratio of the previous three years. Similarly, monthly demand can be estimated by the previous three years' mean monthly sale.

Growth Rate Method

The trend extension method is the application of the historical rate to future growth. If the fertilizer demand during the last five years has grown at an average rate of 7 per cent, it may be reasonably assumed to grow at the same rate for the next year. Projected growth in demand can be modified for the likely impact of new factors, such as changes in produce prices, additional credit availability, and increased area under irrigation or seeds of new varieties. This method is useful for short-term forecasting of fertilizer demand at the country level and of sales of fertilizer companies.

Causal Model Method

The causal model employs the cause and effect relationship between fertilizer demand and the factors that influence it. Important factors influencing demand include fertilizer prices, crop prices, credit availability, irrigated area, rainfall, area under high yielding varieties, cropping pattern and fertilizer distribution arrangements. By analyzing past data, two or three critical factors that have the most profound effect on fertilizer consumption are selected and the effect of the selected factors is quantified and expressed in the form of mathematical equations. For assessment of the demand for future years, the likely state of each selected critical factor at that point of time has to be first assessed.

This method is extremely complex, involving mathematical equations to express the relationships and interrelationships among variables. The reliability of the forecasts, however, depends on values of the selected critical factors and there is no certainty that those set of circumstances will prevail in the years under forecast.

A prerequisite for using this method is the availability of data, by area and season, on various critical factors (fertilizer prices, crop prices, credit availability, area under high yielding varieties, rainfall, etc.) in addition to the data on fertilizer demand for several years. Estimating the influence of each factor on fertilizer demand is a complex statistical process. Its application is not only complex but requires elaborate data for the past which is rarely available. For these reasons, this method is not of practical use either to governments or to marketing organizations.

Lead Indicator Method

This method is meant primarily for short-term demand forecasts. It uses certain lead indicators which need not necessarily have a direct relationship with fertilizer demand. In the case of fertilizer, a possible lead indicator might be the price of produce in the previous season. The higher the price, the more money in the hands of farmers, thus, raising the demand for fertilizer for the following crop. The sale of high yielding seeds prior to planting is possibly another indicator. Precise information on these indicators is not often available and, if available, the relationship between the indicators and fertilizer consumption is rarely consistent and predictable.

Survey of Buyers' Intentions Method

For directly ascertaining the farmers' intentions regarding fertilizer usage, sample surveys of appropriate size are difficult to conduct. If the sample is small, a large degree of error will occur.

Alternatively, it is possible to assess farmers' intentions through fertilizer retailers. Many retailers are themselves practising farmers and have a good assessment of agricultural prospects for the coming season or year. They have a good assessment of produce price trends, available purchasing power, likely crop shifts, etc. It should be possible to ascertain from each retailer directly or through a large sample, the likely fertilizer demand. Through enquiry from their own outlets, marketing organizations can produce a fairly accurate sales forecast.

This method is suitable for short- and medium-term sales forecasts for marketing organizations. The results can be fairly accurate and realistic. The value of a retailer survey lies in almost complete coverage, especially in smaller regions or countries. The cost is low and the survey can be carried out through a company's own sales force.

For forecasting national-level, short-term demand, the retailer sample survey can be employed with advantage as the size of the sample is more reliable than sample survey.

Market Share Method

The market share method arrives at a sales forecast after reviewing marketing plans and assessing the attainable market share. This method is applied when the company wants to know the total sales of fertilizers in its area of operation. As the first step, it should identify its competitors and estimate their sales. It should also know the marketing strategies used by the competitors. It may be possible to ascertain the quantity of fertilizers received at various central warehouses by each competitor.

The market share of the company is determined and it is applied to forecast the fertilizer demand for the next season or year to calculate the company's sales forecast for each province by fertilizer type. It is obvious that this method has to be done in conjunction with a demand forecasting method. Alternatively, the demand forecast by the government or trade association can be adopted as the basis.

The market share method is simple and practical for forecasting the short-term sales. However, this approach has limited relevance when the fertilizer supply is evenly balanced against demand, when each company's sales volume is determined by its ability to procure and supply. This method is thus applicable only when several organizations compete for fertilizer supply in the same region and when supply is abundant.

WTO IMPLICATIONS FOR FERTILIZER MARKETING

The World Trade Organization (WTO) is an international organization established on 1 January 1995 to supervise and liberalize international trade. The WTO is the successor to the General Agreement on Tariffs and Trade (GATT), which was created in 1947. The WTO deals with the rules of trade between nations at a near-global level; it is responsible for negotiating and implementing new trade agreements. The aim of the WTO is to promote free trade and stimulate economic growth.

In pursuance of the guidelines of the World Trade Organization (WTO), quantitative restrictions on import of fertilizers should be removed by the member countries. Most of the countries impose restrictions like no entry or high import duty on the fertilizer imported from foreign suppliers with intention to promote indigenous fertilizer production. These restrictions may appear beneficial for the national economy but in reality turn out to be counterproductive, if the indigenous supply of fertilizers is less than the actual demand. In India, the tariff rate quota for urea imports has been put on hold. At present, there is no bound rate of duty on urea and Government can impose higher tariff in future. But pegging the duty at high level may not be appropriate because imports of urea will become costly to meet the increasing demand-supply gap that is likely to occur in future. The tariff rate quota option will, therefore, provide the flexibility in importing a certain critical quantity at a lower duty.

The WTO related issues have far reaching consequences and therefore they are under detailed examination by the Government in most of the developing countries.

FERTILIZER PROMOTION

The aim of fertilizer promotion is to develop and sustain a brand image of the product. It is basically an extension activity, which is carried out to provide technical service to farmers for popularizing a given product.

The objectives of the promotional programmes are:

- To explain the product benefits
- To motivate and convert potential buyers into consumers
- To build up confidence of farmers on the brand and the company
- To strengthen brand loyalty through pre- and post-sale service

Sound fertilizer promotion programme has strong technical content, good quality products matching the soil-crop needs of the area and efficient product and information delivery system.

Methods of Fertilizer Promotion

All existing tools, techniques and media mixes are used for promoting fertilizers. Common promotional activities carried out by fertilizer industry are as follows:

1. Field programmes:
   Crop Demonstrations/Trials, Farmers' Meeting, Farmers' Camps, Group Discussion, Crop Seminar, Crop Competition, Field Days/Visits, SAU/Research Station Visits
2. Campaigns:
   Soil Testing, Biofertilizer Use, Seed Treatment, Plant Protection, Rat Control
3. Publicity:
   Hoarding, Wall Painting, Fairs/Festivals, A.V. Aids, Puppet Shows, Newsletters, Leaflets/ Folders/Posters, Mass Media (Radio, T.V., Film Shows)
4. Training of farmers, fertilizer dealers, sale point personnel and service centre staff
5. Participation in farmers' fairs, local exhibitions.
6. Farmer's Service Centres
   Farmer's service centres are developed to supplement the concept of 'Single Window System' where all agricultural inputs including farm guidance and soil test facilities are available under one roof. Many of the fertilizer industry members have established the farmer's service centres. India has almost 620 farmer's service centres spread over 17 states of the country. They have been established and are being maintained by the major fertilizer industries.

Government Programmes

In India and other developing countries, government has played an active role in promoting fertilizer use both directly and indirectly. The following government-sponsored programmes have a component of fertilizer promotion:

- National Food Security Mission on rice, wheat and pulses
- Integrated Cereals Development Programme for rice and wheat
- Accelerated Maize Development Programme
- Intensive Cotton Development Programme
- Sustainable Development of Sugar Cane based Cropping Systems
- Balanced use of Fertilizers and Integrated Nutrient Management Programme
- National Pulses Development Project
- Oilseeds Production Programme
- Diversified Agriculture Support Project

Area development approach is much meaningful in achieving the objectives of agricultural development. Area development programmes are undertaken for the benefit of farmers to increase their crop yields through better management of available resources under situations like dryland, problematic soils, tribal areas, etc. Lack of awareness among the farmers, limited financial resources and non-availability of timely and adequate supply of inputs including credit are some of the major constraints which have been identified for adoption of improved technologies. A well-focused programme dealing with location specific problems and their solutions gives good results. Spectacular rise in fertilizer consumption is achieved in a short span of 3-4 years through concerted and collaborated efforts in the selected areas. Emphasis should be laid on area-specific problem-oriented promotional programmes along with service support. Soil testing, integrated plant nutrition system, wasteland development, watershed management, high-tech agriculture, dealers network, use of information technology, credit card facility, etc. are the components for reorienting promotional strategy.

FERTILIZER PRICES AND SUBSIDIES

Fertilizer Prices

In a free market, the fertilizer price is determined by the demand and supply relation. But totally free market situation does not exist. While the fertilizer manufacturer or importer must get a remunerative price (that is, more than the production and marketing cost) to run the industry or business, the farmers should get the fertilizers at an affordable price.

Price of fertilizers has a direct influence on the fertilizer demand. It is a common observation that with increase in price, the demand declines. Decrease in fertilizer demand has a bearing on agricultural production, particularly food production. The linkage between fertilizer price and food security makes it a crucial issue for the Government which has the powers to decide fertilizer prices.

In order to reconcile the uniform sale price and varying cost of production across manufacturing plants, a Retention Price Scheme (RPS) was introduced in India in late seventies. Under this scheme, if a manufacturer's net realization from the sale of fertilizers at the government controlled rate (the same farm gate price throughout the country) falls short of the retention price, the government pays the difference as subsidy to the fertilizer manufacturers. The retention price is fixed for each fertilizer plant by the government after taking into account the type of feedstock or raw material being used, cost of other inputs and maintenance. The basic rationale of adopting the RPS was that it reduces uncertainty of returns on investment and encourages the fertilizer plants to increase capacity utilization. The prices of fertilizer were kept low so that farmers have the incentive to use more fertilizer which has been the very important purchased input for increasing the productivity of crops. In an endeavour to keep fertilizer prices at the levels lower than the economic cost of domestic production or of imports, the gap between the two widened. This gap is what is usually termed as fertilizer subsidy.

Since July 1991 when a programme of economic reforms was launched in India, there have been considerable changes in the fertilizer pricing policy. The low analysis nitrogen fertilizer (like ammonium sulphate and calcium ammonium nitrate) were decontrolled with effect from July 1991. These were again brought under price control a year later but were finally decontrolled with effect from June 10, 1994. Phosphate, potash, NP and NPK fertilizers were decontrolled with effect from August 25, 1992. Thus, beginning from June 1994, all fertilizers, except urea, have been decontrolled. Though the objective of decontrol was to reduce the burden of subsidies on the government exchequer, the actual outgo on this account has not decreased. Owing to the decontrol of phosphate and potash fertilizers, their prices shot up and the government had to announce ad hoc subsidy to reduce the burden on the farmers.

Fertilizer Subsidy

The subsidy on fertilizers is the difference between the production cost and associated costs of fertilizer and the sale price of the fertilizer fixed by the government. In the case of imported fertilizers, the subsidy represents the difference between the CIF (cost, insurance and freight) price of imported fertilizer plus the pool handling charges and the farmer's price (excluding dealer's margins and sales tax). The need of fertilizer subsidy arises when the production cost of the fertilizer is high, which farmers cannot afford to pay. In this situation, the Government fixes the price of the fertilizer at a lower level than the production cost and reimburses the difference to the fertilizer manufacturer.

Subsidy on fertilizers is common in most of the developing countries, though the rate of subsidy varies across countries and types of fertilizers. Though the direct effect of fertilizer subsidy is to provide fertilizer to the farmers at a price lower than its economic price, the ultimate benefits of the subsidy accrue either to the farmers or to the consumers of farm products or both depending on the package of other policies including product price policies. Differential rate of subsidy on nitrogenous, phosphate and potash fertilizers is used to correct the imbalance in fertilizers use.

Subsidy on fertilizers is introduced with the intention that the farmers will get fertilizers at affordable and uniform price. However, increasing cost of production or import and increasing use of fertilizers increases the subsidy burden on the government exchequer. In India, subsidy on fertilizer has risen to huge amount and has become a sensitive and emotive issue. In August 1992, when subsidy on phosphate and potash fertilizers was withdrawn, the price of phosphate fertilizers almost doubled and that of potash fertilizers trebled, consumption of these fertilizers went down and ratio of N : P : K consumption got distorted. Government of India introduced some concession to restore the balanced fertilizer application. Subsidy and concession amounts provided by Government of India has increased to over Rs. 200 billion per annum (Table 12.2). As a long-term policy, the subsidy withdrawal in a phased manner is proposed. But with the withdrawal of subsidy and concessions the prices of fertilizers will increase. In the totally decontrolled scenario, the stability and uniformity of fertilizer prices is not likely to be achieved. In the vast country like India, the farmers who were getting fertilizers almost at the uniform price throughout the country may not continue to avail this facility. They may also witness fluctuating market price of a fertilizer within a short span of one crop season. Such price variation may affect farmers' purchase decision as well.

Table 12.2 Fertilizer subsidy in India

Item	1981	1991	2001	2006
Total fertilizer subsidy (billion Rs.)	5.05	43.89	138.11	204.75
Fertilizer subsidy per farm (Rs.)	7.4	51.9	114.1	162.5
Fertilizer subsidy per hectare (Rs.)	29.2	236.0	726.9	1072
Fertilizer subsidy per tonne NPK (Rs.)	91.5	3497	8,270	10,066
Population (million)	683	846	1027	1095
Farm holdings (million)	89	107	121	126
Gross cropped area (million ha)	173	186	190	191
Fertilizer consumption (million tonne $N + P_2O_5 + K_2O$)	5.52	12.55	16.70	20.34

It is true that fertilizer subsidy exists because fertilizer is sold to the farmers at a price lower than its economic cost. But it is not true that the benefits of this subsidy go only to the farmers who use this fertilizer. Although fertilizer subsidy induces farmers to use more fertilizer which increases the production of agricultural commodities, the benefits of fertilizer subsidy also accrue to the consumers of farm products because as a part of the overall price policy, the prices of farm products are also kept low. The consumers are able to get food products at affordable prices and the industries, which use raw material from the agricultural sector, are able to keep the cost of production of such manufactured goods low.

Selected References

Fertiliser Marketing News, A monthly journal. Fertiliser Association of India, New Delhi.

Jagdish Prasad (2005). Marketing of Farm Inputs: Seed, Fertilizer and Irrigation. In: *Encyclopaedia of Agricultural Marketing IX.* Mittal Publ. New Delhi, pp. 336.

Tandon, H.L.S. (Ed). (1992). *Fertiliser Marketing Systems in India.* Fertilizer Development and Consultation Organization, New Delhi, pp.134.

CHAPTER

13

Fertilizer Logistics

Logistics involves inventory, transportation, warehousing, material-handling, and packaging. Logistics is a channel of the supply chain of goods and services from producer to the consumer. Logistics management is that part of the supply chain which implements the efficient forward and reverse flow and storage of goods, services and related information between the point of origin and the point of consumption in order to meet customers' requirements.

Fertilizer logistics play an important role in marketing of fertilizers and timely availability of the desired product in good condition to the consumers. Storage and transportation of fertilizers assume great significance because of the fact that production of fertilizers is a continuous process whereas fertilizer use by the farmers is seasonal and they do not have proper storage facilities for the fertilizers. Improper storage and handling of fertilizers may lead to loss of nutrients, accidents and environmental pollution.

THE STORAGE OF FERTILIZERS

The fertilizers are handled at two levels: at the production level and at the consumer level. At the production level in the factory, the fertilizer is handled in bulk, while in the marketing stage they are sold in bags. Many of the fertilizers pose serious problems in storage due to their nature and characteristics, particularly, hygroscopic, acidic or explosive nature, under varying climatic conditions. The quantity and quality of fertilizers are liable to be adversely affected in case of defective storage.

The fertilizer requirements of farmers are seasonal unlike its production and import. Moreover, farmers in developing countries do not have suitable storage facilities for fertilizers. Therefore, there should be sufficient storage facilities both for the indigenously manufactured as well as imported fertilizers at the factory site, ports and at strategic consumer centres. In order to overcome transportation problems and to ensure timely availability of fertilizers to the farmers, fertilizers should be stored at consuming centres well in advance in suitable warehouses. In inaccessible locations, fertilizers may have to be stored even for longer periods.

Fertilizer factories have generally storage capacity for about 45 days of their production. It is estimated that 20 per cent of the fertilizer requirements should be kept as buffer stock while 10 per cent of the storage might be at wholesale and retail levels.

The fertilizer storage depends on the following factors:

1. Nature of fertilizer
2. Weather—humidity and temperature
3. Kind of packaging material
4. Type of storage structure
5. Dunnage
6. Stacking

***Nature of fertilizer*:** Some fertilizers are hygroscopic as they pick up moisture easily from the moist air, e.g. urea, ammonium nitrate, CAN, nitrophosphates. By absorption of moisture the fertilizer becomes pasty mass or semifluid in consistency and oozes out of the bag. Moisture also induces chemical changes and the quality of some fertilizers may deteriorate. Moist fertilizer loses free-flowing characteristic and on drying it turns into lumps.

***Weather*:** An increase in the mean temperature of the atmosphere improves the keeping quality of the fertilizer by lowering the moisture content. On the other hand, an increase in the relative humidity of the atmosphere gives rise to rapid deterioration of its keeping quality by an increase in the moisture content of the fertilizer. Relative humidity levels above 60 per cent are harmful and above 80 per cent lead to a rapid deterioration. When humidity is high, the fertilizer depending upon its nature absorbs moisture forming water films around the fertilizer granules. When dry days follow, some moisture is lost and the fertilizer granules stick to one another forming lumps. A sequence of alternating periods of high and low relative humidity results in the formation of bigger and harder lumps.

***Packaging materials*:** The main considerations for selecting a suitable packaging material for fertilizers are tensile strength, resistance to chemicals, impermeability to moisture, shelf-life and ease in filling, handling and storage. Price and resale value are also important economic considerations.

Hygroscopic fertilizers are generally packed in HDPE bags. Polythene laminated jute bags can also be used. For handling convenience, the maximum weight of fertilizer package should not exceed 50 kg so that it can be handled without using iron hooks.

***Kind of storage structure*:** Godowns for storage should protect the fertilizers from high atmospheric humidity and high temperature. In fertilizer manufacturing plants or in big warehouses, the fertilizer can be best stored in bulk. Bulk storage structures with dehumidification facilities are ideal. At some fertilizer factories parabolic RCC bulk storage structures with air-conditioning facilities having storage capacity varying from 25,000 to 1,00,000 metric tonnes in single hall are provided, besides bag storage godowns. For bag storage of hygroscopic fertilizers, the capacity of each room in a godown should not exceed 200 metric tonnes. The godown should be well constructed with moisture-proof flooring with 0.5 or 1 metre high plinth and the storing masonry walls plastered with cement and lime. The roof should be flat RCC. The door and ventilators should be of such type that these can be made airtight when required. Some of the nitrate and sulphates salts have corrosive action on concrete. It is therefore necessary to use protective coating like fluosilicate, sodium silicate, linseed oil or special paints.

Dunnage: Dunnage is a term used for the floor cover for stacking the fertilizer bags. Suitable dunnage is essential to protect the fertilizers from moisture damage from floor as diurnal variation in temperature causes condensation of moisture. Wooden crates dunnage with a layer of matting over it is ideal since it keeps the stock 9-12 cm above the floor level. This will also help free air circulation under the stack and also prevent spillings and droppings of fertilizers coming in direct contact with the floor. In addition, it would also save the fertilizer from water damage in case of any accidental overflow of gutters or leakage. The wooden crates are 1.5 metres × 0.6 metre size and made out of four 1.5 metres lengthwise and four 0.6 metre crosswise battens of 6.5 cm × 6.5 cm section which are uniformly spaced and nailed to make the frame rigid. The timber should preferably be of coniferous trees which would not require painting of solignum.

Other dunnage materials are polythene laminated hessian or a layer of polythene sheet of 300 gauge sandwiched between two layers of bamboo or similar matting. In no case, fertilizer bags should be stacked without dunnage. Even for short period of storage, a minimum of two layers of matting should be used.

Stacking: Proper stacking is a basic requirement for scientific storage. Stack plan should be prepared well before the fertilizer stock. The floor space is divided into uniformly sized and serially numbered rectangular or square stack bases to build stacks. The stack line of 6 cm drawn in white or black paint will serve to mark the boundary for different stacks.

An alleyway of 75 cm between stacks and walls or pillars should be provided. The haulage alleyway should not be less than one metre wide. The size of stack may be variable depending on the quantity of fertilizer and area of the floor. Ordinarily, the maximum base area of a stack does not exceed 9 metres × 6 metres.

There are two methods of stacking: block stacking and cross stacking. In block stacking each layer has tiers of lengthwise and breadthwise bags to form blocks and in any two adjacent layers this system of lengthwise and breadthwise tiers is reversed. Block stacking may be preferred for short periods, while cross stacking for long period of storage. In cross stacking the layers of bags are arranged crosswise; the upper layer has breadthwise arrangement of bags if the lower layer has lengthwise arrangement of bags.

The bags in a stack should be of the same quantity of fertilizer and of uniform size and weight. It is important to see that the stitched mouth of the bags should not face the peripheral side of the stack and no bag should bulge outwards or inwards at any point. Bags on the upper side of the stack should be at the same level. Damaged, cut or torn bags should be kept separately.

The height of the stack is important in fertilizer storage. Fertilizers that are highly hygroscopic, acid treated or explosive in nature should not be stacked over 3-4 metres height while fertilizers that are less hygroscopic may be stacked up to five metres. Examples of these two categories of fertilizers are listed in Table 13.1.

Different fertilizers should be stacked separately. No other industrial or agricultural commodity meant for human or animal consumption should be stored in the same room with fertilizers. Hygroscopic fertilizer stacks should be covered with airtight polythene sheet, tarpaulin or gunny pauls during humid days.

Table 13.1 Stacking height for different fertilizers

Maximum stacking height	*Type of fertilizer*	*Examples*	*Nature of fertilizer*
4 metres	Hygroscopic, acid treated or explosive	Calcium ammonium nitrate	Neutral, explosive
		Ammonium sulphate nitrate	Acidic
		Nitrophosphate	Acidic
		Urea	Hygroscopic
		Single superphosphate	Acidic
		Triple superphosphate	Acidic
		Rock phosphate	May cause damage to gunnies
		DAP	May cause damage to gunnies
		Potassium nitrate	Explosive
5 metres	Non-hygroscopic	Potassium chloride	Neutral
		Potassium sulphate	Neutral
		Ammonium sulphate	Acidic
		Ammonium chloride	Acidic
		Ammonium phosphate	Acidic
		Bone meal	Acidic

WAREHOUSE MANAGEMENT

Warehousing standards are related to the type of product stored. Storage of fertilizers is different from the storage of food and fast moving consumer goods (FMCG) items. Storage of fertilizers can set off chemical reactions, and is likely to damage the flooring of the warehouses. Filled fertilizer bags can be stacked one over the other in a warehouse till permissible heights, but FMCG items which are delicate in nature and have to be dispatched in small amounts, need to be properly stored in racks.

Handling the consignments and unloading, storage and re-dispatch thus require development of efficient processes, which can be easily offloaded to clearing and forwarding (C&F) agents. Normally, the manufacturing unit's enterprise resource planning (ERP) system is extended to the warehouse to integrate and facilitate operations through the C&F agent.

There are three basic functions of warehousing: movement, storage and information transfer. Out of these, movement function generally receives greater attention as organizations focus on improving inventory turns and speeding orders from manufacturing to final delivery.

In case the outward delivery function is hampered due to non-availability of trucks, warehouse function would suffer a major setback.

The movement function normally has the following activities:

1. Receiving the consignment
2. Transfer or put away
3. Order picking/selection
4. Cross-docking
5. Shipping

The receiving activity includes the actual unloading of products from the transportation carrier, the updating of warehouse inventory records, inspection for damage, and verification of the merchandise count against orders and shipping records.

Transfer or put away involves the physical movement of the product into the warehouse for storage, movement to areas for specialized services such as consolidation, and movement to outbound shipment.

Customer order selection or order picking is the major movement activity and involves regrouping products into the assortments of customers' desire. Packing slips are made up at this point.

Cross-docking bypasses the storage activity by transferring items directly from the receiving dock to the shipping dock. A pure cross-docking operation would avoid put away, storage and order picking. Information transfer would become paramount because shipments require close coordination.

Cross-docking has become commonplace in warehousing because of its impact on costs and customer service. For example, most of the food distribution involves the cross-docking of products from supplier to retail food stores. Eliminating the transfer or putaway of products reduces costs and the time goods remain at the warehouse, thus improving customer service levels.

Shipping, the last movement activity, consists of product staging and physically moving the assembled orders onto carrier equipment, adjusting inventory records, and checking orders to be shipped. It can consist of sorting and packaging of items for specific customers. Fertilizer bags are placed in trucks or rail wagons and are marked with information necessary for shipment, such as origin, destination, shipper, consignee and package contents.

Information transfer, the third major function of warehousing, occurs simultaneously with the movement and storage functions. The manufacturer always needs timely and accurate information as it attempts to administer the warehousing activity. Organizations are relying increasingly on computerized information transfer utilizing electronic data interchange (EDI) and bar coding to improve the speed and accuracy of the information transfer. Trained staff reporting to the C&F agent ensures that the information flow happens with accuracy.

PACKAGING

In developing countries common fertilizers are solid materials, which are packed in 50 kg jute or HDPE bags. Smaller packages of expensive fertilizers are also marketed. For example, DAP is packed in both 50 and 40 kg bags, whereas micronutrients fertilizers are usually packed in 5 or 10 kg bags. Packings of 5, 10, 20 or 25 kg are also marketed for hilly or inaccessible areas or to meet the needs of special projects.

Each fertilizer bag bears the name of fertilizer, net weight, nutrient content and name of the producer or importer firm. This is the requirement as per legislation in India and other countries. Brand name of the fertilizer, if any, is also printed on the bag to make it appear different from similar products or as quality assurance.

MOVEMENT OF FERTILIZERS

Fertilizer movement and its consumption require transporting and storage at a large number of rural centres before reaching its final destination, i.e. millions of farmers spread over vast area of the country. Fertilizer movement has to be timely to cope with the vagaries of monsoon, seasonality of demand as also the spread out rail and roads.

Fertilizer distribution in large countries like India require maximum input of transportation as the fertilizer has to be transported and distributed among small landholders spread in villages throughout the country. Another reason which provides a pivotal role of transportation in the area of fertilizer availability is localized nature of fertilizer plants and ports on one hand and widespread distribution of consumers over a large area.

Good infrastructure and effective logistics management are essential to ensure that the fertilizer demand is converted into sales. While farmers may actually need and are able to buy fertilizer, this demand cannot materialize unless fertilizer is transported and stored with wholesalers and retailers in time for seasonal application. In most developing countries, availability of fertilizer and its proper distribution are factors severely limiting fertilizer use. Efficient transport can bring about an immediate increase in fertilizer consumption.

There are three modes of transport for fertilizers: rail, road and inland water transport. During 2006-07, the railways moved 79% of the fertilizer produced and imported in India, road transport carried 20% fertilizers while inland water transport carried merely 0.5% compared to 10% in China and 11% in USA. It has been estimated that one horsepower can transfer 40,000 kg material by waterways, 500 kg by railways and 150 kg by road. Thus, waterways provide the cheapest means for transport.

The effectiveness of a mode of transport is expressed in terms of lead, which is the distance over which the material moves. The judicious management of the demand-supply balance helps to reduce the average lead of fertilizer movement by rail. Since 1980-81, the overall lead of fertilizer movement by rail has progressively come down from 1100 km to 804 km during 2006-07.

Roads provide transportation for primary movement of fertilizers directly from factories to ports to farm gate. So far as secondary and tertiary short lead cartage is concerned, road is the only available means of transport. However, road transport is not economical to move fertilizers over long lead of greater than 250 km. Road condition determines the speed of movement. Good roads save not only time and money but also ensure timely supply of fertilizers to farmers.

Road transport is convenient in the sense that the goods can be transported quickly at a short notice and the consignment reaches the warehouse directly whereas the rail transport has to be planned much in advance and it carries the consignment upto the railway station only. Also, roads are the only means to carry the materials to inaccessible areas.

Handling of Fertilizers

The solid fertilizers are handled many times during their transfer from the factory to the farm. The number of times a fertilizer is handled depends on three factors:

1. Physical form in which the fertilizer is transported: Whether it is transferred in bulk or in bags?
2. Distribution system: Whether the fertilizer distribution system is 2-tier or 3-tier system?
3. Transport system: Whether it is transported by rail, road or waterways?

A fertilizer that is imported in bags is handled 16 times by the time it reaches the farm gate. The fertilizers which are produced indigenously are handled 8-10 times during their transfer from the factory to the farm. The repeated handlings of the bags have an adverse effect on their condition. The damage of the bag may cause spilling of the material during transfers leading to loss in the quantity of the fertilizer by the time it reaches its destination.

The work of loading and unloading of trucks and rail wagons is performed by porters who are habitual of using iron hooks. Use of iron hooks in fertilizer bags is strictly prohibited because it causes many problems such as:

1. Use of iron hook makes a hole in the bag through which fertilizer granules or particles spill over.
2. After use of hook, the fertilizer particles fall over other bags, absorb moisture from the air and react with the bag. This makes the bag weaker and prone to damage. The jute bags are more vulnerable to damage than the HDPE bags.
3. The holes made by the hooks allow entry of moist air into the fertilizer bag causing caking. In hygroscopic fertilizers this becomes a serious problem.
4. The spilled over granules or particles make the floor of the carrier and the warehouse dirty.

Handling of fertilizer bags by mechanical devices is safer than their manual handling. In large-scale factories, mechanical methods of material transfer are used. But availability of sufficient manpower in India attracts manual handling which needs careful supervision.

Selected References

Fertiliser Marketing News, A monthly journal. Fertiliser Association of India, New Delhi.

Tandon, H.L.S. (Ed). (1992). *Fertiliser Marketing Systems in India.* Fertilizer Development and Consultation Organization, New Delhi, pp.134.

CHAPTER

14

Fertilizer Application

Fertilizers are readily available sources of nutrients and they are used for supplementing the nutrient supply of the soil for providing adequate quantity of each nutrient to the plants. The utilization of the nutrients supplied through fertilizers by the crops depends largely on the method and time of fertilizer application. Other factors that influence the utilization of fertilizers by the crops are crop characteristics, soil characteristics and climate.

Developments in fertilizer technology have provided an array of commercial fertilizer types and formulations, including solids, liquids, gases, suspensions; single-elements to complex multi-nutrients; and granules and powder forms. Research in soil chemistry and plant nutrition has provided the technical basis for fertilizer application methods.

The broad goals of fertilizer application are:

- cost-effective production of quality crops;
- efficient use of fertilizers;
- maintenance of soil quality; and
- protection of the environment.

This chapter briefly highlights the fertilizer application methods within such a context.

METHODS OF FERTILIZER APPLICATION

Common methods of application of the solid fertilizers are broadcast application, band placement and foliar fertilization.

BROADCAST APPLICATION

In broadcast method, the fertilizer is applied uniformly over the field either before planting or after planting. Broadcast applications usually involve large amounts of fertilizers and lime. In preplant broadcast application the fertilizer is incorporated with the soil by tilling or cultivating. Incorporation with the soil increases the crop recovery of immobile nutrients like phosphorus (P) and potassium (K), while rainfall or irrigation can move the mobile nutrients like nitrogen (N) into the root zone without incorporation. Broadcasting is the common method of application of solid fertilizers in both the basal dressing as well as topdressing. Application of high rates of the major nutrients, namely; nitrogen, phosphorus and potassium, is usually carried out by broadcasting manually or with a tractor-mounted fertilizer spreader. This allows a specific amount of fertilizer to be spread over the entire area. For nitrogen fertilizers which are soluble materials, broadcast method is efficient in conventional tillage systems but in zero-till systems this method can greatly reduce the nutrient recovery by the crop due to poor access by plant roots and losses through volatilization, denitrification and immobilization.

Phosphate and potash fertilizers are invariably applied as basal dressing as these two nutrient elements are little mobile in the soil, so they must be mixed with the soil in the root zone. The efficiency of P and K fertilizers applied by broadcast method even though mixed with soil is relatively low due to poor access by the plant roots and interaction with soil constituents particularly clays, iron oxides and free lime.

Some advantages of broadcast application of fertilizers are:

1. An application of large amounts of fertilizers is made without danger of injury to the plant.
2. If tilled into the soil, distribution of nutrients throughout the plough layer encourages deeper rooting and improved exploration of the soil for water and nutrients.
3. Labour is saved during planting.
4. This method can be a practical means of applying maintenance fertilizer, especially in zero-till cropping systems.

Uniform and accurate spreading of fertilizers is essential for effective utilization by the crop. The effects of uneven application of the NPK fertilizers on a low fertility soil results in yield loss. There may also be yield reductions due to over-fertilization on medium or high fertility soils. The effects of under-fertilization and over-fertilization of N on small grains are especially prominent.

BAND PLACEMENT

Band placement refers to the placement of fertilizer as a band either on the surface or below the soil surface close to the seed or to the side of the plant roots. Fertilizer placement can be done as soil surface application or subsurface application before planting, at planting or after planting. The placement method depends on the crop, soil fertility status, mobility of the nutrient in the soil and equipment availability.

Preplant Placement

Surface band placement: Surface band or dribble-applied fertilizers can be an effective method of preplant application. However, if not incorporated, dry surface soil conditions can reduce nutrient uptake, especially with immobile nutrients such as phosphorus. Surface band application of nitrogen can improve N availability to crops compared to broadcast application in some soils and cropping systems.

Subsurface band placement: Subsurface banding is effective in increasing fertilizer efficiency. The depth of placement varies between 5 and 20 cm depending on the crop and the fertilizer material. Subsurface "point or spoke injection" can be effective for immobile nutrients like P and K. In zero-till system, point injection of N is also more efficient than broadcast method.

At Planting

Subsurface band: Fertilizer placement can be made at many locations near the seed, depending on the equipment and crop. Solid as well as fluid fertilizers can be placed as subsurface band. However, subsurface application of fluid fertilizers is more common. The fertilizer is placed 2.5 to 5 cm below the seed or 2.5 to 7.5 cm to the side and below the seed depending on the equipment.

Seed band: Fertilizer application with the seed is commonly used as starter or pop-up application to enhance early seedling vigour, especially in cold, wet soils. This method is not used in tropical environment. Usually low rates of fertilizers are applied to avoid germination or seedling damage. Fluid or solid sources can be used.

Surface band: Fertilizers can be surface applied or dribbled at planting in bands directly over the row or several centimetres to the side of the row. Application over the row can be an effective method of placement of immobile nutrients (P and K) with a hoe type opener because soil can "slough off" over time and cover up the fertilizer band. Thus, the surface applied band becomes a subsurface band placed slightly above the seed.

After Planting

Top dressing: This refers to broadcast application in the standing crop. Top dressing of N fertilizers is common in small grains, vegetables, pastures and other seasonal crops. Phosphate fertilizers are not effective as top dressing. Recent researches show that top dressing of potash fertilizers in irrigated crops is as effective as basal dressing.

Side dressing: Side dressing of N fertilizers is very common with cereals, maize, sorghum, cotton and other crops in USA and other developed countries. This is done by an equipment having standard knife or point injector applicator. Anhydrous ammonia and fluid N sources are most commonly applied by this method. Side dressing allows more flexibility in application time since it can be made anytime the equipment can move in the field without damage to the crop.

Side dressing of P and K fertilizers is not recommended because most crops need P and K early in the season.

Surface band: Fluid fertilizers can be surface band applied or dribbled beside the row after planting.

Application of Specific Nutrients

***Nitrogen**:* In early seedling period, only small amount of N is required by the crops and the soil has some amount of readily available N which can meet this requirement. But agronomists have been making the conventional recommendation of heavy application—about 50-75% of the total N fertilizer requirement of the crop, to be applied as basal dressing probably because it is more convenient to apply N as preplant basal dressing, rather than top dressing in the standing crop. For irrigated and favourable rainfed environments, this perception is now changing in view of the following:

1. The utilization of fertilizer N applied as preplant basal dressing is low compared with mid-season or late application because the roots development in early stages is inadequate to absorb the applied N. The highest recovery of fertilizer N in rice has been observed prior to heading stage. The same is true for other grain crops.
2. The unused fertilizer N applied as preplant basal dressing has strong potential for environmental pollution. The nitrate is either leached down to groundwater or it is subjected to denitrification producing nitrous oxide (N_2O) which is a greenhouse gas and also causes ozone depletion.
3. Fertilizer N applied as topdress in mid-season or in late season is easily spread in root zone as water movement carries the N down to the plant roots.

For dry land agriculture, however, basal dressing of N is more important because utilization of topdress N is uncertain if surface soil dries out.

In irrigated crops N fertilizer is topdressed in 2-3 split doses and the time of top dressing is tailored to the irrigations. However, N fertilizer application must end before flowering of grain crops.

Phosphorus: Since phosphate is immobile in the soil, its placement near roots is advantageous. Topdressing of phosphate by broadcasting in the standing crop will not place the P in the root zone and it will not be accessible to the plant roots of the crop in the season. Phosphate is therefore invariably applied as basal dressing. A notable exception is the forage crops which can utilize top dressed phosphate effectively.

Generally, band placement of phosphate is considered superior to broadcast but this is not always true. It has been observed that relative efficiency of band placed or broadcast phosphate is related to both the P status of the soil and the rate of application. As the P rate increases, broadcast application can be equal to banding. Also the difference between band placed and broadcast P application decline with increasing levels of available soil P. Thus, in low P soils it is better to go for banding while in medium or high P soils broadcast application may be made.

Utilization of fertilizer P by the first crop is usually low not exceeding 25% even with band placement. This is in marked contrast to N and K which show usually a crop recovery of 50 and 75%, respectively. Band placement of P reduces contact with the soil and results in less fixation than broadcast application. When soluble P fertilizer is broadcast and mixed with soil, fertilizer P comes into intimate contact with large volume of soil. It is therefore expected that band placement of P will increase the crop recovery as compared to broadcast P.

The unused P remains in the soil, though in altered form, which is less available than the original fertilizer P. This is the source of P build up in the well-fertilized intensive cropping systems.

Potassium: Potassium salts are soluble and mobile in the soil but much less than nitrate (NO_3^-) and more than phosphate ($H_2PO_4^-$). Generally, K fertilizers are band placed with P fertilizers as preplant application. However, K fertilizer salts should not be mixed with the seed or placed in contact with seed in great quantity. They should be placed in a band to the side and below the seed.

Broadcast application of K fertilizers is usually less efficient than band placement. However, the difference between the placement methods decreases with increasing soil K status. Similarly, the importance of placement decreases as higher rates of fertilizer are used.

Recent work on irrigated rice and wheat has shown that part of the K (upto 50%) can be topdressed like N. The topdressed K is effectively utilized by the crop as the K can move down with water to the zone of root activity and will be absorbed. In dry land system K should be applied as basal dressing.

Secondary nutrients: Calcium and magnesium are constituents of liming materials which are applied in large quantities as broadcast. Magnesium deficiency in crops is sometimes reported which requires broadcast application of sulphate of potash-magnesia. Sulphur fertilizers are also broadcast with NPK fertilizers or they may be banded near the seed. Elemental sulphur, ammonium polysulphide, pyrites and ammonium thiosulphate are acid-forming fertilizers which lower down the soil pH around the granule and correct the micronutrient (zinc, iron, manganese, copper) deficiencies that are induced by high pH values.

Micronutrients: Micronutrient deficiencies are increasing and are likely to appear in more crops over large areas in future. Higher yields are being harvested which is putting a greater demand on all nutrients. The secondary and micronutrients are thus being mined continuously from the soil leading to development of deficiencies. Micronutrients may be added to N-P-K fertilizers or may be applied separately as a broadcast application or as a foliar spray.

Micronutrients added to N-P-K fertilizer should be placed in bands about 5 cm away from the seed to prevent injury. Boron should not be band applied to sensitive crops like beans and small grains, which are easily injured by boron toxicity.

FOLIAR FERTILIZATION

Foliar fertilization refers to spraying of growing plants with suitable fertilizers solutions. These solutions are prepared in low concentration to supply single plant nutrient or a combination of nutrients. It has been established that all plant nutrients are absorbed through the leaves of plants and absorption is remarkably rapid for some nutrients. Foliar nutrition has several potential benefits, such as:

- Supplying nutrients during peak periods of demand when an immediate response is needed.
- Providing plants with certain nutrients, such as zinc and iron, that may not be readily available for root uptake.
- Allowing flexibility in supplying nutrients related to improving the quality of the harvest.
- Controlling nutrient losses in conditions with high potential loss.
- Providing a nutrient source during periods of stress when soil applications are not practical.
- Giving a nutritional boost to plants at the same time when other foliar chemicals are being applied, thereby minimizing application expenses.

Foliar feeding is one of the most efficient ways of supplying essential nutrients to a growing crop. It provides a method of delivering essential nutrients at key growth stages when the plant can utilize them the most.

The nutrients are taken into the foliage and transported to all parts of the plant in a short period of time. Soil-applied nutrients depend upon moisture to dissolve the fertilizer, and much of those nutrients are tied up in the soil.

Though it is theoretically possible to feed a plant completely via the leaves, in practice, so many sprays would be required that the process would be rendered uneconomical.

Foliar fertilization is not a substitute for soil fertilization but an effective additional means of supplying sudden demands. Foliar spraying of fertilizers is costly compared to soil application unless combined with other spraying operations taken up for insect or disease control. Though foliar application does not result in a great saving, it may be favoured under following conditions:

- When visual symptoms of nutrient deficiencies are observed during early stages of deficiency.
- Unfavourable physical and chemical conditions of soil reduce fertilizer use efficiency.
- During flooding and drought conditions wherein soil application could not be done due to excess or for want of moisture, foliar application of soluble fertilizers is taken up.
- Under very high and uncontrollable weed population conditions, foliar application of soluble fertilizers is the only option.
- Foliar fertilizer application can be made at the time of spraying plants with plant protection chemicals.

Certain difficulties are associated with foliar application of nutrients as detailed below:

- Marginal leaf burn or scorching may occur if higher doses of soluble fertilizer solution are used.
- As solutions of low concentrations (usually 3 to 6 per cent) are to be used only a small quantities of nutrients can be applied in a single day.
- Several applications are needed for moderate to high fertilizer rates.

Urea is the most common N source that is used in foliar sprays. Plants differ in their tolerance of urea concentration. Generally, grain crops can tolerate higher concentration than the vegetable, fruits and ornamentals. With use of high pressure low volume sprayer, higher concentration of urea can be sprayed. Safe concentrations of urea for spraying on various crops are given in Table 14.1.

Table 14.1 Safe concentrations of urea for spraying on various crops

Crop	*Urea concentration (%)*	
	Low pressure high volume sprayer (Ordinary sprayer)	*High pressure low volume sprayer (Micronet sprayer)*
Wheat	3–6	10–30
Rice	2–4	10–20
Maize	2–4	10–20
Cotton	1–2	10–20
Sugar cane	4–6	15–30
Vegetables	0.5–1.5	5–10
Fruits	0.2–1.0	3–5
Ornamental plants	0.5	1–2

A portion of a plant's nutritional needs can be met by applying soluble fertilizer directly to the foliage. Foliar fertilization can result in rapid nutrient absorption and utilization to correct deficiencies or to merely prevent nutrient shortages during critical periods of growth. However, unlike roots, plant leaves are not adapted to assimilate large amounts of nutrients and meet the bulk of the nutrient requirement.

For some crops, foliar nutrition may be the most economical and reliable method of providing some nutrients, especially with micronutrients. However, there can be a large difference in the effectiveness of various fertilizer sources in actually penetrating into the leaf surface and providing the desired nutritional benefit. A local expert should be consulted to select the source of foliar nutrition that will best achieve the desired result with the least expense on specific crops.

There are many environmental factors that also impact the effectiveness of foliar nutrient application. Generally, application during early morning or evening is most effective. Air temperatures less than 30ºC and high humidity conditions favor nutrient absorption into the leaf. Wind speed should be low to avoid missing the target plants and minimize drift. Nutrient application to young, actively growing tissue and buds is generally more effective than application to mature tissue.

Foliar application of macronutrients, such as N and K, can also be beneficial for meeting the plant nutritional requirement. For example, foliar application of K has been shown to help cotton meet the demand of the rapidly developing bolls when roots may not be capable of completely meeting the demand. Additionally, foliar sprays of N fertilizer onto small grains such as wheat are sometimes beneficial in increasing the protein content of the grain.

Foliar fertilization can provide an important supplement to the nutritional programme of farmers. However, this practice should be considered an additional management technique and not the primary means of nutrient delivery. Plant roots have evolved to be the major pathway for nutrient uptake and their health and function is the primary goal. Appropriate soil fertility levels in the root zone should be maintained through monitoring of the fertility status by regular soil testing and nutrient applications. However, during critical growth stages, a foliar application of nutrients might provide the plants what they need to achieve goals for yield and quality.

An appropriate time to consider foliar fertilization is when a specific nutrient shortage is evident based on visual symptoms or plant analysis. If a nutrient deficiency exists, then foliar application would be one means of providing a quick but temporary fix to the problem. Certain soil conditions such as high pH, low pH, drought, excessive moisture, or cool temperatures may cause some nutrients to be unavailable for uptake by the roots. If any one of these conditions exists, the problem may be more effectively corrected with foliar application than with soil application.

TIME OF FERTILIZER APPLICATION

Fertilizer application timing depends on the crop, nutrient, soil and climate. Ideally, fertilizer applications should be tailored to match the nutrient needs of the crop during the growing season to maximize the nutrient use efficiency and minimize the environmental pollution potential. In practice, however, it is more the convenience and economics that determine the timing of fertilizer application, rather than the scientific principles underlying efficient fertilization of crops. Nutrientwise information about fertilizer application timing is discussed below:

***Nitrogen**:* Nitrogen is highly mobile nutrient which does not stay for long at the point of application. This nutrient is required by plants throughout the growing season from early vegetative growth till reproductive phase with peak demand near heading stage in grain crops. In irrigated crops like wheat, rice and sugar cane the total amount is split into 3-4 instalments. For example, in irrigated wheat 33-50% of the total N is applied as basal dressing and the remaining amount is divided into three instalments at crown root initiation (first irrigation), active tillering (second irrigation) and jointing (third irrigation). In transplanted rice, basal N dressing is small (20 kg/ha) and the major portion is applied in 3-4 splits at early tillering, active tillering and panicle initiation (40-45 days after transplanting in medium duration cultivars).

Under unirrigated conditions fertilizer use is low and major portion is applied as basal dressing. The topdressing is given only after rainfall occurs.

Phosphorus: Phosphate fertilizers are generally applied either just before or at planting because this nutrient is required by crops during early growth period and it is immobile in the soil. To minimize the fixation of soluble phosphate in the soil, the phosphate fertilizer should be banded as discussed in the previous section. Topdressing of phosphate is not practised as it is not an effective method of application.

Potassium: Like phosphate, potash fertilizers are generally applied and incorporated before or at planting. Usually side dressing is less efficient than broadcast. Top dressing of potash fertilizer before heading is also effective in irrigated grain crops.

***Secondary nutrients**:* Usually calcium and magnesium are supplied to acid soils through lime, which is broadcast and mixed several weeks before planting. For correction of specific nutrient deficiency in crops like fruit trees in mid-season, the individual nutrient can be applied as foliar spray but soil application should be done to prevent appearance of the deficiency in the next season.

Sulphur sources are also applied as preplant broadcast and mixed with soil. However, it is advisable to apply elemental sulphur and pyrites one or two weeks before planting to allow their oxidation and conversion into sulphate form which is used by the crop. For correction of sulphur deficiency in crops in the mid-season soluble sulphur sources such as gypsum and potassium sulphate can be applied as foliar spray but soil application should be done to prevent appearance of the deficiency in the next season.

Micronutrients: Micronutrients sources are applied as preplant broadcast or band placement near the seed. Soil application of iron salts is however not effective. For correction of micronutrient deficiency in crops in the mid-season soluble sources are applied as foliar spray but soil application should be done to prevent appearance of the deficiency in the next season.

Fertigation

Fertigation refers to the application of nutrients through an irrigation system. It provides a simple and effective way to supply nutrients, particularly nitrogen and potassium. Most drip (trickle), jet, spray and sprinkler systems can be used. But fertigation is not effective using flood or furrow irrigation as the distribution of nutrients will be uneven and leaching will occur.

Fertigation has the potential to ensure that the right combination of water and nutrients is available in the root zone, satisfying temporal requirement of the plants. Fertigation saves fertilizer as it permits applying fertilizer in small quantities at a time matching with the plants nutrient need. Besides, it is considered eco-friendly as it avoids leaching of fertilizers.

Fertigation is used extensively in commercial agriculture and horticulture and is starting to be used in general landscape applications as dispenser units become more reliable and easy to use. In Israel about 80% of the irrigated land uses the fertigation method, combining the application of water and fertilizers through the drip irrigation system. The production of vegetables, flowers, ornamental plants and spices in greenhouses has been experiencing accelerated growth in developed countries in recent years and most of the greenhouses are computerized, allowing automatic control of water, fertilizers and climate systems.

Fertigation allows an accurate and uniform application of nutrients to the wetted area, where the active roots are concentrated. Therefore, it is possible to adjust the nutrients quantity and concentration to their demand through the growing season of the crop. Recommendations have been developed for the most suitable fertilizer formulation (including the major nutrients, NPK and microelements) according to the type of soil, physiological stage, climate and other factors. Special attention should be given to the pH and NO_3^-/NH_4^+ ratio, nutrient mobility in soil and salinity conditions.

Since fertigation ensures nutrient supply to the crops according to their physiological stage of development, considering the soil and climate characteristics, it gives high yields and high quality crops with minimum pollution of the environment.

In orchards, fertigation has the advantage of supplying nutrients to the area of greatest root activity, that is, the irrigated part of the root zone. Nutrients can be applied when the trees need them with little wastage. This is particularly important on sandy soils with low nutrient retention capacities. Fertigation allows regular top-up of nutrient reserves.

Depending on the irrigation system being used, applications of fertilizer may need to be split over several irrigations. This also improves the percentage taken up and utilized by the tree. Frequent applications are easy to apply and there is no soil compaction problem as would be the case with broadcasting with a tractor. Careful fertigation can reduce the annual requirement of nitrogen and potassium due to a reduction in losses to leaching and weed competition, resulting in cost savings and easier weed control.

AMOUNT OF FERTILIZER TO APPLY

The key question in fertilizer use for any crop is: how much fertilizer should be applied? This question has not only economical significance for a farmer but it also has national and environmental significance. While applying less than the required amount of fertilizers will not produce the optimum crop yield, applying more than the required amount will not only be wasteful but can also be detrimental to the yield and quality of the crop and cause environmental pollution.

During the last six decades, intensive research in soil-plant relationship has led to development of scientifically sound principles of quantifying the crop needs of nutrients. Depending on the availability of basic information needed for making fertilizer recommendation for a crop, the following methods are used for deciding the quantity of the fertilizer nutrients needed by a crop to be grown on a soil.

General recommendation: This is also referred to as "standard agronomic recommendation". The fertilizer doses for the crops in different agro-climatic zones are estimated from the crop response functions derived from the data for simple fertilizer trials conducted at few locations in the area. The general fertilizer recommendation does not take into account site-to-site variation in nutrient supply capacity of soils to meet crop demand. Thus, the fertilizer recommendations are crop specific but not soil specific. The same recommendations are made for a large area, regardless of the type of soils and variation in their fertility and other properties. This method is widely used in developing countries where soil testing service is not developed and fertilizers are relatively inexpensive. This method may allow over-fertilization of the crop in some fertile fields while providing inadequate amount of fertilizers for the same crop on a poor fertility soil. The net result would be loss of production and poor fertilizer use efficiency.

Soil test based fertilizer recommendation: A soil test is a rapid chemical method for estimating the nutrient supplying power of a soil. A soil test actually measures part of the total nutrient supply in the soil and represents only an index of nutrient availability. Soil tests do not measure the exact quantity of a nutrient that will be taken up by a crop. To predict the nutrient needs of crops, the soil test must be calibrated against nutrient rate experiments in the field.

Usually, a soil test is done in a well-equipped Soil Testing Laboratory run by the Government or private organizations. For getting a soil tested, representative composite soil sample from the plough layer (0-15 cm depth) of the field following the standard procedure is submitted to the Soil Testing Laboratory. As the soil testing is done several weeks before the planting of the crop, it provides valuable data on soil fertility status which forms the basis for formulating fertilizer recommendations for crops on different soils. Soil-test based fertilizer recommendation is site-specific as the fertilizer recommendation for the same crop may be different for different soils, in contrast to uniform standard agronomic recommendation. As the soil test values become lower, the probability of obtaining a profitable response to fertilization becomes greater (Fig. 14.1).

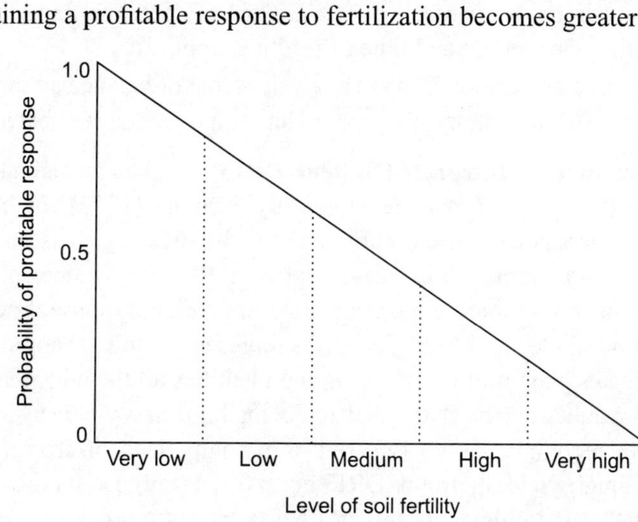

Fig. 14.1 Level of soil fertility determines the profitable response to fertilization

The accuracy of soil test based fertilizer recommendation depends on the amount of basic information available for making the recommendation. The basic information relates to the total nutrient requirement of the crop for optimum yield, efficiency of measured soil available nutrient and efficiency of fertilizer nutrients for the crop. These cardinal values change with the crop, cultivar, season, soil and crop management practices and nature of the fertilizer. In the beginning of the development of soil test based fertilizer recommendation sufficient data are not available for accurate prediction of fertilizer needs of crops on each type of soil. In this situation, soil test values are grouped into 'low', 'medium' and 'high' categories and fertilizer recommendations for the soil samples are made category-wise. Subsequently, when sufficient basic data become available soil test based fertilizer recommendations for targeted yield of crops are made by "prescription method". The latter allows consideration of the soil fertility status of individual soil sample and fertilizer use efficiency in the soil.

Recently, online fertilizer recommendations based on soil test values have become available in India and other countries.

Soil test based fertilizer recommendation is also called site-specific nutrient management as it duly considers the soil variability among fields and this forms a component of precision agriculture.

Plant Analysis Based Fertilizer Recommendation

Plant analysis based fertilizer application is commonly used for commercial and horticultural crops. In this method some plant part, usually most recently matured leaf, is selected and analyzed for all the mineral elements. The nutrient concentration is interpreted for sufficiency level using critical nutrient concentration or range. For example, critical nutrient concentrations of N, P and K for maize (in the leaf opposite and below the uppermost ear) at silking time are 3% N, 0.5% P and 2% K, respectively. Fertilizer applications are advised if the particular nutrient is found insufficient, below the critical nutrient concentration or range.

An excellent example of the use of plant analysis for crop nutrition is the crop logging carried out for sugarcane in Hawaii. For orchards, high value crops and greenhouse crops plant analysis is widely used for estimating the amount and time of fertilizer application.

One of the problems in interpretation of plant analysis is that of balance among nutrients. Ratios of nutrients in the plant tissue are commonly used to determine the nutrient balance in crops.

Diagnosis and Recommendation Integrated System (DRIS): The Diagnosis and Recommendation Integrated System (DRIS) approach was developed by Beaufils (1973) for interpreting results of foliar analysis for diagnosis of nutrient sufficiency and deficiency. It is an alternative to the critical nutrient concentration approach discussed above. DRIS uses indices of ratios of nutrient concentrations and has been found to be more accurate in predicting nutrient needs for numerous crops than the critical nutrient level system. The DRIS represents a holistic approach to interpreting tissue analyses. It evaluates plant nutritional status and identifies all the nutritional factors limiting crop production. Thus, it enhances the chances of achieving high crop yields by improving fertilizer recommendation. DRIS recommendation is based on a comparison of crop nutrient ratios with optimum values from a high-yielding group (DRIS norms). The comparison gives the *index values* that measure how far the particular nutrients are from the optimum level and are used in the calibration to classify the yield factors in order of limiting importance.

DRIS has several advantages over the critical level approach in making diagnosis for fertilizer recommendation:

1. The importance of nutritional balance is taken into consideration in deriving the norms and making diagnoses.
2. The norms for the nutrient content in leaves can be universally applied to the particular crop, regardless of where it is grown.
3. Diagnosis can be made over a wide range of crop development, irrespective of the cultivar.
4. The nutrients limiting yield through either excess or insufficiency can be readily identified and arranged in order of limiting importance for yield production.

RECENT APPROACHES FOR IMPROVING FERTILIZER EFFICIENCY

Real-time Nitrogen Management

It is now well recognized that plant uptake of applied fertilizer nitrogen holds the key to fertilizer efficiency. The crop plants use nitrogen as per their need and not necessarily in direct proportion to the N supply. Matching N supply with crop demand will optimize N use efficiency and reduce N losses to the environment. Crop demand of N varies with growth stage. It is low in the early stage and increases with growth reaching maximum near heading. For achieving higher efficiency fertilizer N application should therefore be synchronized with actual crop conditions and nutrient requirements. Synchronization of fertilizer nitrogen application with crop N demand can result in high yields, reduced losses of N and efficient utilization of applied N, avoiding excessive use of fertilizer. Adjusting fertilizer application to crop N needs during growth period is the basic principle of real-time N management.

Real-time N management requires quick and reliable tools to decide on the time when fertilizer N needs to be applied to the crop. It also requires the identification of an optimal leaf N content or leaf colour that must be maintained throughout the season to obtain high yields. This is done with the help of a diagnostic tool to monitor the plant N status in the field and calibration of leaf N status with crop growth and yield. There are four diagnostic methods to assess the plant N status:

a. Total N determination of the leaves
b. Chlorophyll meter
c. Leaf colour chart
d. Optical sensors

Total N determination of the plant leaves is most reliable but most expensive and time consuming. It can therefore be used for only high value crops or fruits and vegetables. It is not suitable for large-scale adoption in cereal crops.

The chlorophyll meter has been used to monitor plant N status *in situ* in the field and to determine the right time of N topdressing in rice, maize and wheat. By using this tool, one can synchronize fertilizer N application with actual crop demand. The chlorophyll meter (or SPAD meter) is a simple, portable diagnostic tool that measures the greenness or relative chlorophyll content of leaves (Inada, 1963, 1985; Kariya et al., 1982). But the chlorophyll meter is too expensive (Rs. 70,000 per unit) for farmers in developing countries.

The leaf colour chart (LCC) is an alternative tool for the expensive chlorophyll meter for real-time or crop-need-based N management in crops. It is simple, easy-to-use and inexpensive. It costs about Rs. 30 per piece, which farmers can afford easily. It helps farmers to determine the right time of N application. It measures leaf colour intensity that is related to leaf N status. LCC is an effective tool to optimize N use in cereals like rice, wheat and maize at high yield levels, irrespective of the source of N applied, viz. organic manure, biologically fixed N, or chemical fertilizers.

Fig. 14.2 Monitoring leaf nitrogen status by leaf colour chart

Use of leaf colour chart for nitrogen fertilizer scheduling of rice has been successfully demonstrated in several Asian countries including India, China, Indonesia, Thailand, Vietnam and Philippines by the International Rice Institute, Manila, Philippines. Critical LCC values for different rice varieties and crop establishment methods are shown in Table 14.2.

Table 14.2 Critical leaf colour chart (LCC) values for different rice varieties and crop establishment methods

Rice variety	*Crop establishment*	*Critical LCC value*
Scented and aromatic	Transplanted	2
Semidwarf indica	Direct-seeded	3
Semidwarf indica	Transplanted	3.5
Hybrid	Transplanted	3.5

Rice crop with leaf colour below the threshold greenness shade suffers from N deficiency and requires immediate N fertilizer application to prevent yield loss. In 24 on-farm trials on LCC-based nitrogen management in rice carried out in north-western India, rice yields were similar to those obtained by applying recommended dose of 120 kg N ha^{-1} but about 30 kg N ha^{-1} less fertilizer was used by real-time nitrogen management. In a similar study around Pantnagar (Uttarakhand, India) 30-60 kg fertilizer N ha^{-1} in irrigated rice as well as wheat was saved and higher crop yields were recorded with use of LCC in comparison to farmers' fertilizer practice.

Optical sensors for use at farm level have recently been developed for guiding N topdressing in cereals and turf grass. Sensor based fertilizer recommendations are available online.

Site-specific Nutrient Management

Soil test based fertilizer application in rice is not very successful for a variety of reasons including flooding of the soil which brings about drastic changes in soil chemical, physical and biological properties. As an alternate to the soil test based fertilizer recommendation, the site-specific nutrient management (SSNM) approach was developed in rice-producing Asian countries through partnerships of the Irrigated Rice Research Consortium (IRRC) under aegis of the International Rice Research Institute, Manila, Philippines. This approach emphasizes 'feeding' the crop with nutrients as and when needed. This involves dynamic adjustment of fertilizer application to optimally fill the deficit between the nutrient needs of a high-yielding crop and the nutrient supply from naturally occurring indigenous sources such as soil, irrigation water, manures, organic amendments and crop residues. The SSNM approach aims to apply nutrients at optimal rates and optimal times to achieve high yield and high efficiency of fertilizer use by the crop.

The SSNM approach has recently been simplified for large-scale adoption by the farmers. There are *four* essential steps involved in this approach for estimating fertilizer N, P, and K requirements of the crop.

1. **Establish an attainable yield target for farmers' fields**

   The yield target is directly used to calculate fertilizer rates for the crop. It must be realistic and attainable by farmers. A yield target higher or lower than one attainable by farmers would lead to recommendations of more or less fertilizer than required for high fertilizer use efficiency and profit. Grain yield from a fully fertilized plot with no nutrient limitations and good management can be used to estimate the yield target.

2. **Approximate a fertilizer N rate and formulate dynamic N management**

   The difference between the yield target and N-limited yield (yield with no N fertilizer) provides an estimate of anticipated crop response to fertilizer N. The estimated yield response to fertilizer N and a targeted efficiency for the fertilizer N are used to approximate the total requirement of the crop for fertilizer N, which is dynamically apportioned among 3-4 applications to best match the crop's need for N. For this, leaf colour chart is used as inexpensive diagnostic tool to monitor leaf N status which helps in deciding the right amount and time of N fertilizer top dressing.

3. **Estimate field-specific nutrient-limited yields**

   Nutrient-limited yields are determined by the nutrient omission plot technique. The P-limited yield is determined in a P omission plot receiving no P fertilizer but sufficient N and K to ensure they do not limit yield. The K-limited yield is determined in a plot receiving no K fertilizer but sufficient supply of other nutrients.

4. **Determine fertilizer P and K rate**
   The crop's need for fertilizer P is based on a comparison of the yield target and P-limited yield, whereas the crop's need for fertilizer K is based on a comparison of the yield target and K-limited yield. The SSNM approach advocates sufficient use of fertilizer P and K to overcome P and K deficiencies and maintain soil P and K fertility.

The benefits of SSNM approach for rice production have been successfully demonstrated in Asian countries by the Irrigated Rice Research Consortium under aegis of International Rice Research Institute, Manila, Philippines. In this method, fertilizers are applied to rice using the following principles to achieve high yield and high efficiency of plant use:

1. Apply only a moderate amount of fertilizer N to young rice within 14 days after transplanting (DAT) or 21 days after sowing (DAS), when the need of the crop for supplemental N is small.
2. Apply fertilizer N after 14 DAT or 21 DAS at active tillering and panicle initiation stages based on the crop's need for supplemental N, as determined by leaf N status. The leaf colour chart (LCC) is a tool that could be used for assessing leaf N status and the crop's need for N.
3. Apply all fertilizer P near transplanting or sowing.
4. Apply fertilizer K twice: 50% near transplanting or sowing and 50% at early panicle initiation. When fertilizer K rates are relatively low (for example, ≤ 30 kg K_2O ha^{-1}), all fertilizer K can be applied near transplanting or sowing.

SSNM approach has also been successfully used in wheat, maize and other crops. But it is not yet being used on a large scale mainly because the development of SSNM recommendation is time consuming and highly technical.

Integrated Plant Nutrient Management

The basic concept of integrated plant nutrient management (IPNM) is the adjustment of plant nutrient supply to an optimum level for sustaining the desired crop productivity through optimization of the benefits from all possible sources of plant nutrients in an integrated manner. It involves appropriate combination of chemical fertilizers, organic manures, crop residues, N_2-fixing crops and biofertilizers according to the system of land use and ecological, social and economic conditions. IPNM takes a holistic view of soil fertility and plant nutrition management for a targeted yield of crops in the cropping system. It aims at sustainable agriculture with reduced dependence on fertilizers and increased use of local resources.

The integrated plant nutrient management thus firstly operates at plot level, optimizing the utilization of plant nutrients from diverse sources which are locally available, in order to improve the agronomic efficiency of such nutrients and at the same time reducing the losses of nutrients. Organic recycling is specially promoted through IPNM for facilitating nutrient recycling to improve soil fertility and productivity at low cost.

Components of IPNM

The main components of IPNM are chemical fertilizers, organic manures including green manure and agro-industrial by-products, biofertilizers, crop residues and leguminous crops. The relative contribution of each component to IPNM for a cropping system or farming system depends mainly on their local availability and socio-economic factors. Use of locally available alternative sources of nutrients particularly organic materials is certainly encouraged. It is recognized in IPNM concept that no single source of nutrient, be it chemical fertilizer, organic manure, green manure, biofertilizers or crop residue, is good enough to meet the full nutritional requirements of an intensive cropping system.

IPNM focuses on cropping system rather than an individual crop and the farming system rather than an individual field to ensure sustainable production. The IPNM practices are economically justified, socially acceptable and eco-friendly.

IPNM also aims at making nutrient uptake more efficient. Most crops make inefficient use of nitrogen, often losing nitrogen into the atmosphere through volatilization. New techniques, such as deep placement of fertilizers and the use of inhibitors or urea coatings, have been developed to address this problem. With these innovations, and with better timing and more concentrated fertilizers, nutrient uptake efficiency can be expected to improve by as much as 30 per cent in the developed world and 20 per cent in developing countries by the year 2020.

The steps involved in IPNM are:

- Assessment of on-farm and off-farm resource availability through survey related to soil and nutrient management.
- Fixing yield target depending on the resource availability.
- Soil test based estimation of nutrient requirement with due consideration on soil amendments.
- Integration of all nutrient resources available for a given circumstances.
- To determine time, method, mode of application considering the type of crops involved in the cropping system.
- To adopt efficient soil and water conservation measures to check soil erosion, soil organic carbon and nutrient losses.
- Maintaining soil fertility in terms of soil physical, chemical and biological properties and processes.

Use of Remote Sensing and Precision Agriculture

Remote sensing techniques offer a unique solution for mapping the growth and nitrogen status of crops and monitoring its time course. Therefore, these maps are powerful tools for recommending spatially precise nitrogen application.

Use of new tools to prescribe nitrogen (N) fertilization rates based on wheat canopy reflectance is promising in areas where rainfall is certain during late tillering and jointing stage. The use of canopy spectral reflectance indexes (normalized difference vegetative index, NDVI) using an optical sensor like Green Seeker has been successfully used to prescribe N fertilization in several wheat areas around the world.

Selected References

Havlin, J.L., Beaton, J.D., Tisdale, S.L. and Nelson, W.L. (2005). *Soil Fertility and Fertilizers*, 7th ed. Prentice-Hall of India, New Delhi.

Mishra, B., Shekhar, K.S., Bharadwaj, A.K. and Buresh, R.J. (2005). Site-Specific Nutrient Management for Sustainable Rice-Wheat Production. *Res. Bulletin* **149**, G. B. Pant Univ. Agri. & Tech., Pantnagar. p. 31.

Mordtvet, J.J., Murphy, L.S. and Follett, R.F.H. (1999). *Fertilizer Technology and Application*. Meister Publishing Company, Euclid Ave., Willoughby, OH.

Ryan, J. (2006). Fertilizer application methods. In: *Encyclopaedia of Soil Science* (R. Lal, Ed.), Taylor and Francis, London/New York.

CHAPTER

15

Long-term Effects of Fertilizer Use

Long-term agricultural field experiments provide valuable information about sustainability of different farming practices, as measured by yield trends and changes in soil quality over decades. They provide information on the impact of long-term use of fertilizers and nutrient management systems with varying sources, types and combinations of plant nutrient inputs on soil fertility and productivity. Such knowledge is crucial for any decision support system which aims to develop strategies for enhancing and sustaining agricultural productivity.

At present there are around 600 long-term experiments running in the world, with duration of more than 20 years, 25 of them continuing for more than 100 years.

The world's oldest long-term agricultural field experiments were begun between 1843 and 1856 by Lawes and Gilbert at Rothamsted in England to learn how crop management techniques influenced the yield and soil quality over time. Eight of these experiments still continue. Similar experiments started all over Europe, for example, in France at Grignon in 1875, in Germany at Bad Lauchstädt in 1902, in Denmark at Askov in 1894, in Russia at Moscow in 1912. In the USA, there are four classical long-term experiments: Morrow Plots at the Univerity of Illinois (1875), Sanborn Field at University of Missouri (1888), Magruder Plots at Oklahoma State University (1892) and Old Rotation at Auburn University (1896). All were started as agronomic experiments to determine the nutrient requirements of agricultural crops. Although the original questions have long since been answered, they continue to provide data of considerable agronomic, ecological, environmental and scientific value. Many of these experiments still continue today and provide valuable data on the impact of different fertilizing strategies on soil properties and crop yield.

The Rothamsted Long-Term Experiments

The Rothamsted long-term field experiments in England were started over 150 years ago. These experiments were originally designed to investigate the mechanism by which crops acquired nitrogen. The first harvest in 1844 of winter wheat experiment showed nitrogen to be the nutrient that gave greatest increase in yield. The continuity of these experiments has provided the data for assessing the sustainability of agricultural practices. Although the questions the experiments were originally designed to answer have long been resolved, the experiments continue to give results of interest to agronomists, ecologists, soil scientists, plant pathologists, and others. The experiments show that grain yields can be sustained (and even increased) for over 150 years in monocultures of wheat and barley given organic *or* inorganic fertilizer annually. They provide data on the long-term effects of inorganic fertilizers and organic manures on soil organic matter levels. These data have been used to test computer-based models for the turnover of organic matter in the soil.

Long-Term Fertilizer Experiments in India

In the beginning of 20th century, based on Rothamsted model, a series of long-term fertilizer experiments were established at different locations in India. These were at Kanpur (Uttar Pradesh) in 1905, Pusa (Bihar) in 1908 and Coimbatore (Tamil Nadu) in 1909. Later on more long-term fertilizer experiments were established on heavily fertilized crops such as sugar cane at

- Shahjahanpur (Uttar Pradesh) in 1935
- Padegaon (Maharashtra) in 1939
- Indore (Madhya Pradesh) in 1947
- Muzaffarnagar (Uttar Pradesh) in 1949 and
- Anaka Palle (Andhra Pradesh) in 1950

An experiment on cereal-based cropping system was started at Ranchi (then in Bihar, now in Jharkhand) in 1956.

Unfortunately, some of these experiments were either discontinued or seriously altered as they were found inadequate in respect of the statistical requirement pertaining to the design of experiments or suffered from some management problems. However, the trials at Coimbatore and Ranchi are still continuing.

With the advent of green revolution in 1968, use of fertilizers and other inputs increased substantially to meet the nutrient requirements of high yielding and fertilizer responsive crop varieties. With this arose the need to monitor the changes in soil fertility and soil quality under intensive farming for developing a database for the decision support system on agricultural practices. This called for long-term studies at selected sites for monitoring the changes in the soil fertility and soil quality for developing strategies for intervention. The Indian Council of Agricultural Research (ICAR), in collaboration with State Agricultural Universities initiated the All India Coordinated Research Project on Long-Term Fertilizer Experiments (LTFE) in 1972 at 11 centres located in major agro-climatic zones of the country. These were located at Bangalore, Barrackpore, Bhubaneswar, Coimbatore, Delhi, Hyderabad, Jabalpur, Ludhiana, Palampur, Pantnagar and Ranchi. Six more centres—one each at Akola, Junagadh, Parbhani, Pattambi, Raipur and Udaipur were initiated during 1995-96 to represent additional agro-climatic regions. The project has the coordination cell at Indian Institute of Soil Science, Bhopal.

Effect of Fertilizers on Crop Yield

Edmeades (2003) reviewed the results from 14 field experiments comparing the long-term (20 to 120 years) effects of fertilizers and manures (farmyard manure, slurry, and green manure) on crop production and soil properties. The input of nutrients as either fertilizers or manures had very large effects (150-1000%) on soil productivity as measured by crop yields. The long-term experiments confirm that without any fertilizer application, crop yields decrease to stabilize at relatively low levels. As seen from Table 15.1 in the long-term experiment on maize-wheat at Palampur, use of N fertilizer alone gave double the yield obtained in the unfertilized control in the beginning but the yield declined after 10 years due to deficiency of other nutrients.

Table 15.1 Grain yield (t/ha) of wheat during 1973-1996 (5 years moving average) in the long-term fertilizer experiment at Palampur

Fertilizer treatment	*Period*				
	1973–1977	1978–1982	1983–1987	1988–1992	1992–1996
Control (No fertilizer)	0.70	0.24	0.24	0.29	0.35
100% N	1.39	0.41	0.17	0.13	0.05
100% NP	2.80	1.70	1.72	1.68	1.41
100% NPK	2.99	2.03	2.81	2.79	1.98
100% NPK + FYM	3.73	2.40	3.74	3.57	2.99
100% NPK + Lime	3.32	2.62	3.34	3.29	2.72

Source: Sharma *et al.* (1998)

The NPK also could not maintain the initial level of productivity due to either inadequacy of the nutrient application or deficiency of other nutrients or both. Regular addition of FYM with NPK enhanced the wheat yield by 51% over NPK alone. It may be concluded that the long-term use of fertilizers is sustainable only when they are applied in sufficient and balanced amount in conjunction with organic manures to fulfil the nutrient requirements of crops.

Impact on Soil Organic Matter

Fertilizers generally increase soil organic matter because the increased crop growth returns larger amounts of residues to the soil. In the All India Coordinated Research Project on Long-term Fertilizer Experiments, use of NPK in different intensive cropping systems for 22 years resulted in substantial increase in the organic carbon content over the initial level at 9 out of 11 locations representing different soil types and agro-ecological regions in India (Swarup, 1998). At two locations—Pantnagar and Ranchi—there was a decline in organic carbon level over years in the NPK plots as compared to the initial values. The application of FYM in conjunction with NPK could however maintain the soil organic carbon level at the initial level at Ranchi under soybean-wheat and at Pantnagar under rice-wheat, whereas marked increase in the soil organic matter content was recorded at all other locations. The increase in soil organic matter in NPK and NPK + FYM plots is conspicuous in view of the fact that all the straw had been removed from the plots. In another long-term experiment on maize-wheat at Ranchi, a marked increase in the soil organic matter content after 45 years of cropping with use of fertilizers was recorded (Table 15.2).

Table 15.2 Effect of 45 years (1956-2000) of nutrient use in maize-wheat on soil fertility at Ranchi

Treatment	*Soil pH*	*Organic C, %*	*Available P*	*Available K*
Initial	5.5	0.52	7.5	360.0
Unmanured	5.5	0.64	9.2	75.5
N	4.0	0.58	18.9	91.2
NP	4.0	0.72	248.2	68.5
NPK	4.1	0.83	255.2	72.9
FYM	6.2	1.46	37.5	94.9
NPK + Lime	6.5	0.55	94.7	75.3

Source: Sarkar and Singh, 2002

At Pantnagar, a drastic decline in organic matter level in N, NP and NPK plots has been observed (Table 15.3), though the fertilized plots maintained much higher level of organic matter than the unfertilized control. It is noteworthy that the Pantnagar soil was initially very rich in organic matter having been brought under cultivation recently after deforestation of the area and it is likely that the soil organic matter had not attained equilibrium with the climate under agricultural production system.

Table 15.3 Effect of long-term fertilizer use in rice-wheat for 31 years (1971-2002) on available nutrients of the soil

Treatment	pH (1 : 2.5)	EC (dS m^{-1})	*Organic carbon* (%)	*Available* N[@] (kg ha^{-1})	*Available* P[$] (kg ha^{-1})	*Available* K[*] (kg ha^{-1})
100% NPK	7.7	0.35	0.84	231.4	18.5	124.0
150% NPK	7.8	0.38	0.87	310.0	30.8	155.0
100% NPK + Zn	7.8	0.34	0.88	226.0	18.3	124.7
100% NP + Zn	7.8	0.32	0.88	235.0	17.9	98.3
100% N + Zn	7.7	0.30	0.80	232.0	7.2	93.7
100% NPK + FYM	7.8	0.37	1.54	293.6	28.8	144.7
Control	7.8	0.29	0.49	175.0	6.1	97.0
Initial value	7.3	0.35	1.48	392.0	18.0	125.0

[@] Alkaline permanganate method, [$] Olsen-P, * Neutral ammonium acetate method

Source: Sharma *et al.* (2006)

Build-up and Depletion of Nutrients

It is well known that use of nitrogen fertilizer does not result in accumulation of inorganic nitrogen (ammonium-N + nitrate-N) in the soil because the nitrate form of inorganic N which is the end product of N transformation in the soil is ephemeral and is easily lost through leaching or denitrification. That is why, N fertilizers are known to have no residual effect on the next crop, in contrast to P and K fertilizers.

The soil nitrogen is directly linked with organic matter. A build-up of organic matter is reflected in increase in total N content. As the organic matter content increases with long-term fertilizer use, so should the N supply. Organic manures exhibit distinct impact on organic matter and nitrogen content in the soil.

As regards the effect of phosphate fertilizers, only small fraction (15-25%) of the applied P is usually utilized by the crop receiving this nutrient. Since phosphate is immobile in the soil, it remains fixed or locked up in the soil, if not used by the crop. This results in build-up of P in soil over years if sufficient amount of P fertilizer is applied every season. This trend is clearly visible from the data of long-term fertilizer experiments (Tables 15.2 and 15.3). On the other hand, if only N fertilizer is applied without P addition, a decline in soil P level is a certainty leading to decline in crop productivity.

In the All India Coordinated Research Project on Long-term Fertilizer Experiments, application of NP and NPK in different cropping systems for 22 years resulted in substantial build-up of P over the initial level at all locations representing different soil types and agro-ecological regions except at Coimbatore (Swarup, 1998).

In contrast to phosphate, potassium requirement of crops is high (sometimes greater than N), yet low rates of potash fertilizers are applied. In this situation the available K level in the soil declines gradually with intensive cropping irrespective of nutrient input systems followed. For this reason, decline in available K level in the soils has become widespread in India. The results of long-term fertilizer experiments vividly show this trend. In the All India Coordinated Research Project on Long-term Fertilizer Experiments, application of NK and NPK in different cropping systems for 22 years resulted in a build up of available K over the initial level at 7 out of 11 locations representing different soil types and agro-ecological regions (Swarup, 1998). There was a marked decline in available K level in the soil at Coimbatore, Hyderabad, Jabalpur and Pantnagar indicating that the crop removal of K was greater than the K added through fertilizer.

Soil Acidification

Soil acidification is a process by which soil pH decreases over time. There are no visible signs of the problem, apart from a decline in crop yields if the soil pH drops to very low levels (below 5.0). Soils can acidify under natural conditions over thousands of years, with high rainfall areas most affected. However, they can also acidify rapidly over a few years under intensive agricultural practices.

Soil behaves like a buffer that resists change in the pH. The buffer action of soil depends largely on the soil texture and organic matter content. The nature of clay minerals also determines the extent to which the soil is buffered. This property of the soil enables it to maintain its reaction (pH) and prevent abrupt change in soil pH. Soil acidification happens when the acidity produced by acid-forming compounds such as elemental sulphur and nitrogen fertilizers exceed the buffering capacity of the soil to which they are added. Resulting changes to the soil's chemical and physical properties can have negative impact on vegetative growth and soil health.

Acidic soils may have some or all of the following problems:

- beneficial soil microorganisms may be prevented from recycling nutrients (e.g. nitrogen supply may be reduced)

- phosphorus in the soil may become less available to plants
- deficiencies of calcium, magnesium and molybdenum may occur
- the growth of plant roots may be restricted and their ability to use subsoil moisture may be limited
- aluminium, which is toxic to plants and microorganisms, may be released from the soil
- levels of manganese may reach toxic levels
- uptake by crops and pastures of the heavy metal contaminant, cadmium, may increase.

Factors that accelerate soil acidification include:

- acid rain in industrially developed countries such as Europe and North America
- the application of ammonium-based nitrogen fertilizers to naturally acid soils at rates in excess of plant requirements
- leaching of nitrate nitrogen, originally applied as ammonium-based fertilizers, out of the root zone
- continued removal of alkaline or basic elements through plant and animal produce and waste products from the field.

Acid rain is caused by sulphur dioxide emitted by power plants using fossil fuels and nitrogen oxides emission from automobiles. Natural causes such as lightning, volcanic eruptions and fires also contribute to acid rain. Fertilizers play practically no significant role in acid rain.

The use of fertilizers, particularly nitrogen fertilizers, can accelerate the natural process of soil acidification. The initial effect of fertilizers on soil pH in the immediate vicinity of the granules is usually of little consequence. While some fertilizers are acidic, e.g. ammonium sulphate and others basic, e.g. DAP, their effects are confined to the point of application. Due to the relatively small volume of soil affected compared to total soil volume, there is little effect on soil pH. Some fertilizers, e.g. anhydrous ammonia and urea may initially raise the soil pH at the site of application but in the long term they acidify the soils. This occurs when ammonium is converted to nitrate through the process of nitrification in the soil.

Fertilizer materials vary in their effect on soil reaction (pH). Nitrate sources carrying a basic cation such as calcium and sodium do not acidify the soil. The nitrogen fertilizers containing or forming NH_4-N exhibit greater effect on soil pH (Table 15.4).

Superphosphate is sometimes held to be responsible for soil acidification. It does not directly cause soil pH to fall, but since it does promote legume growth and nitrogen fixation, it will, like most nitrogen fertilizers, have an indirect effect on soil pH. Any practice which increases plant growth such as proper fertilization will lead to the gradual acidification of the soil by two main processes. They are the formation and loss of nitrate through the soil, and removal of alkaline elements such as calcium, magnesium and potassium in the form of farm produce.

Common potash fertilizers such as muriate of potash and sulphate of potash have no permanent effect on soil acidity.

Removal of plant and animal produce removes alkaline cations such as calcium, magnesium and potassium along with some acidic elements. This has an overall acidifying effect on the soil.

Any accumulation of organic matter in the soil increases the amount of nitrogen stored in the soil. Organic nitrogen is then converted to ammonium and subsequently to nitrate by nitrification. In the process, hydrogen ions are formed, i.e. acid is produced. Acid produced in the nitrification process can be used if the nitrate is taken up by plants or soil organisms, but if the nitrate is leached beyond the root zone, acidification occurs. Then the effect of intensive agricultural systems is one of soil acidification.

The estimates of potential acidity of commercial fertilizers are given in Table 15.4.

Table 15.4 Potential acidity of commercial fertilizers

Material	*Grade* (N-P_2O_5-K_2O-S)	*Potential acidity* ($CaCO_3$ *equivalent*), (kg/100 kg *material*)
Anhydrous ammonia	82–0–0–0	140
Urea	46–0–0–0	71
Ammonium chloride	25–0–0–0	136
Ammonium nitrate	34–0–0–0	62
Ammonium sulphate	21–0–0–24	110
Calcium ammonium nitrate	25–0–0–0	Neutral
Monoammonium phosphate	11–52–0–0	58
Diammonium phosphate	18–46–0–0	65
Single superphosphate	0–16–0–13	Neutral
Triple superphosphate	0–46–0–3	Neutral
Rock phosphate	0–32–0–0	Alkaline
Basic slag	0–6–0–0	Alkaline
Nitrogen solution (Urea-ammonium nitrate)	28–0–0–0	50
Elemental sulphur	0–0–0–90	231
Muriate of potash	0–0–60–0	Neutral
Sulphate of potash	0–0–52–17	Neutral
Potassium-magnesium sulphate	0–0–22–11	Neutral
Gypsum	0–0–0–18	Neutral

Use of acid forming fertilizers in small amounts or for a short period does not affect the soil pH. However, continuous use of high doses of acid-forming fertilizers for 10-20 years or longer will certainly result in increased soil acidity. The change appears early (in 10 years) in light texture soils such as sandy loam than in heavy texture soils like clay loam (15 years or more). In India there are reports of increased soil acidity due to continuous use of ammonium sulphate (Table 15.2) while in highly buffered neutral soil of Pantnagar slight rise in soil pH has been observed due to continuous use of urea and other fertilizers for 31 years. It has been observed that the actual increase in soil acidity caused by acid-forming fertilizers such as ammonium sulphate or urea is usually 20 to 50% of the potential acidity of the fertilizer. There are considerable differences among different soils.

The acidifying effects of nitrogen fertilizer can be corrected if lime is applied. In temperate regions, lime is applied in quantities of many tonnes per ha, but less frequently than fertilizers, to provide optimum conditions for growth of many temperate crop species. Apart from neutralizing soil acidity, liming improves the availability of other nutrients such as phosphate, and lowers the toxicity of aluminium and manganese. In the humid tropics, the lime requirements are high and the effect may not last long due to leaching losses. However, increase in crop yield can sometimes be achieved with minimal applications of lime due to alleviation of aluminium toxicity and/or calcium deficiency and care must be taken to avoid over-liming.

Impact of Fertilizers on Soil Structure

Evidence from many long-term experiments indicates that use of mineral fertilizers has no adverse effect on soil structure or physical condition. It is well known that soil physical condition is determined largely by the organic matter content. The long-term use of fertilizers helps build-up of organic matter which enhances the aggregating action. The latter is also enhanced by root proliferation and greater amount of decaying residues from well-fertilized crops making the soils more friable, easier to cultivate and more receptive to water. In a study, Buol and Stokes (1997) concluded that organic carbon contents that become lower under inadequate fertilization recovered when adequate fertilizer was applied. Adequate fertilization also contributes to greater biomass production tending to protect soil from erosion and providing greater quantities of residue critical to soil aggregation.

Field plots at the Rothamsted Experimental Station in England, which have received chemical fertilizers since 1843, are more productive today than at any time in the recorded past. At the Askov experimental station in Denmark, after 90 years the plots receiving NPK fertilizers had an 11% higher organic carbon content than the control plot. The increase in organic matter content induced by NPK applications resulted in a decrease in soil bulk density and a concomitant increase in total porosity.

At Pantnagar continuous use of NPK fertilizers in rice-wheat system for 31 years resulted in lowering of soil bulk density and increase in the porosity, water transmission rate (hydraulic conductivity) and the stability of soil structure as measured by mean weight diameter of soil aggregates (Table 15.5).

Table 15.5 Effect of continuous fertilizer use in rice-wheat for 31 years (1971-2002) on some physical properties of the soil at Pantnagar

Treatment	*Bulk density* (mg m^{-3})	*Hydraulic conductivity* (mm h^{-1})	*Mean weight diameter*[@] (mm)
100% NPK	1.35	0.74	0.72
150% NPK	1.32	0.75	0.72
100% NPK + Zn	1.31	0.76	0.72
100% NP + Zn	1.30	0.76	0.71
100% N + Zn	1.38	0.72	0.68
100% NPK + FYM	1.20	0.80	1.04
Control	1.52	0.68	0.63

Source: Sharma *et al.* (2006) [@] an index of soil structure stability

Impact on Soil Erosion

Soil erosion is a process of land degradation, which is more serious in the regions where a dry season alternates with a season with torrential rains. At the end of the dry season the soils have sparse vegetative cover, particularly if the land has been overgrazed by livestock. Under semi-arid conditions, wind erosion and desertification are serious problems. A fertile soil supporting a swiftly growing crop is much less erosion prone than a poor soil supporting a poor crop. The better developed the surface canopy the greater the protection against wind and water action. Because of their vigorous root systems and large residues, high yielding crops help to anchor the soil. The roots and residues returned to the soil enhance productivity by building up organic matter, by improving aeration and by increasing water infiltration rates. The residual effects of the greater organic matter production are significant also in improving soil aggregation. Land use management appropriate to the topography and rainfall, together with the appropriate use of fertilizers, can make an important contribution to soil conservation. Reduced tillage cultivation practices significantly reduce erosion. The proportion of cropland subject to no-till techniques is expanding rapidly in the USA and certain other countries, for example, in Brazil.

Effect on Soil Quality

Soil quality is the ability of a soil to perform its three primary functions: a basic resource for crop production, to partition and regulate water flow, and to act as an environmental filter. It is a function of many factors, including agro-climatic conditions, hydrology and cropping system and cultural practices. Soil quality is determined by a set of many interrelated physical, chemical and biological properties.

Soil has both inherent and dynamic qualities. Inherent soil quality is a soil's natural ability to function. For example, sandy soil drains faster than clayey soil. Deep soil has more room for roots than soils with hardpan near the surface. These characteristics do not change easily. Dynamic soil quality refers to the properties that change depending on how the soil is managed. It includes the soil pH, amount of organic matter, soil structure, water and nutrient holding capacity. For sustaining soil quality agricultural practices should be such that they improve soil function in the given ecological framework.

Continuous use of fertilizers do not affect the inherent soil quality parameters like texture or mineralogical make-up but it does affect the dynamic soil quality directly or indirectly through changes in soil pH, organic matter content, nutrient status, physical condition, soil erosion and runoff. The long-term effects of fertilizers on the soil quality indicators are summarized in Table 15.6.

The effects on soil quality are positive if the fertilizers are applied judiciously in adequate amount at right time by following appropriate method but may be negative if the fertilizers are used indiscriminately without concern for soil health and environment.

Table 15.6 Long-term effects of fertilizers on soil quality indicators

Soil quality indicator	*Influence on productivity*	*Long-term effect of fertilizers*
Texture	Determines nutrient reserves, friability, air-water relationship, drainage and erodability	No effect
Structure (aggregation)	Determines aggregate stability, air-water relationship, erodability	Balanced fertilization improves soil aggregation
Bulk density	Determines porosity, air-water relationship, reflects impact of compaction	No direct effect, balanced fertilization has favourable effect indirectly
Water-holding capacity	Determines the choice of tillage operations and measures to improve retention and release of water and nutrients	No direct effect; balanced fertilization has favourable effect indirectly
Organic matter	Influences favourably all physical, chemical and biological properties. As a sink to CO_2 of the atmosphere, improves environmental quality	Balanced fertilization enhances root growth which adds organic matter
Cation exchange capacity	Regulation of retention and release of water and nutrients for higher crop productivity	No direct effect; indirectly increases CEC through enrichment of organic matter
Soil reaction	Guidance for reclamation measures to shift pH towards normal range of 6.8 to 7.5	Some nitrogen fertilizers and ammonium phosphates acidify the soil; potash fertilizers and superphosphates have no effect
Available nutrients	Determines the soil fertility status, deficiency or excess of a nutrient	Excessive use of N fertilizers alone causes depletion of other nutrients: P fertilizers help build up of P; low rates of K cannot replenish its mining from the soil
Microbial activity	Determines decomposition rate of organic matter and nutrient release, especially N and S, nitrifying power, nitrogen fixation	Moderate and balanced amount of fertilizers promote microbial activity but excess amount of inorganic N may kill soil microbes and earthworms

Selected References

Abrol, I.P., Bronson, K.F., Duxbury, J.M. and Gupta, R.K. (2000). *Long-term soil fertility experiments with rice-wheat rotations in South Asia*, Paper Series 6, Rice-Wheat Consortium for the Indo-Gangetic Plains, New Delhi.

Nambiar, K.K.M. (1994). *Soil Fertility and Crop Productivity under Long-Term Fertilizer use in India*. Publication and Information Division, ICAR, New Delhi.

Sharma, S.P., Sharma, J. and Subehia, S.K. (1998). Long-term effects of chemical fertilizers on crop yields, nutrient uptake and soil environment in western Himalayan soils. In: *Proc. National Workshop on Long-Term Soil Fertility Management through Integrated Plant Nutrient Supply* (A. Swarup *et al.* Eds). Indian Institute of Soil Science, Bhopal, pp. 125-138.

Singh, G.B. and Swarup, A. (2000). Lessons from long-term fertility experiments. *Fertiliser News* **45** (2): 13-24.

CHAPTER

16

Fertilizers and Environmental Pollution

Environmental consciousness received much impetus after publication of "Silent Springs" by Rachel Carson in 1962. In the last quarter of the 20th century, concern about the impact of fertilizers on environment was expressed by the environmentalists and ecologists. In many areas, nitrate and phosphate enrichment of water bodies was noticed and fertilizers were suspected to be the source of these nutrients. Development of Hypoxia or Dead Zone in the Gulf of Mexico and subsequently mortality of fish (Pfiesteria problem) in Chekaspeake Bay, USA in 1997 highlighted the seriousness of the environmental pollution caused by nutrient enrichment of water bodies. The nutrients—nitrates and phosphates—were believed to come from the run-off from agricultural fields, though they originate not only from the fertilizers but also from soil, animal manures, sewage, septic tanks, detergents, etc.

ENVIRONMENTAL POLLUTION BY FERTILIZERS

Fertilizers produce environmental pollutants both at the factory level during manufacture and at farm level after their use in agriculture. Both at the point of production and when applied to stimulate crop growth, fertilizers have an impact on greenhouse gas emissions. Taken in isolation, fertilizer production leads to inevitable emissions of some greenhouse gases. But these can be minimized, and the fertilizer industry has taken great strides to improve its performance in this area during recent years. Depending on management practices, crop fertilization can either produce a positive, negative or neutral impact on environment.

POLLUTION DUE TO FERTILIZER PRODUCTION

NITROGEN FERTILIZER PLANTS

The nitrogen fertilizer industry is an energy-intensive industry, producing carbon dioxide (CO_2) from both process (feedstock) and fuel (energy) sources. Ammonia plants produce much carbon dioxide which is a greenhouse gas. According to the latest International Fertilizer Association benchmarking exercise, the CO_2 generation from ammonia production ranges from 1.52 to 3.06 tonnes of CO_2 per tonne of ammonia produced. On average, one-third of CO_2 emissions are from burning fuel and two-thirds are from the use of hydrocarbon feedstock. Natural gas is the dominant feedstock for the ammonia plant. If natural gas is used as the feedstock in a modern ammonia plant using steam reforming process to prepare synthesis gas, about 2.7 tonnes of CO_2 per tonne of ammonia-N are produced. The CO_2 production increases 25% when coal or fuel oil is used as feedstock. However, this CO_2 can be utilized for urea production; about 1.6 tonnes of CO_2 per tonne urea-N is required.

Globally, human activity contributes 7% of the CO_2 released to the atmosphere and the share of fertilizer industry is only 2% of that. Thus, the contribution of ammonia plants to CO_2 emission is only 0.1-0.2%.

The production of nitric acid, which is used for manufacture of ammonium nitrate and nitrophosphate, results in emission of nitrous oxide (N_2O) which has very high global warming potential (310 times more than CO_2). Furthermore, N_2O is detrimental to ozone layer and is associated with widening ozone hole in the stratosphere. The rate of N_2O emission varies from 1 to 10 kg per tonne of 100% nitric acid. It is estimated that fertilizer production currently accounts for 6% of the man-made N_2O emission compared with nearly 50% from the motor vehicles.

In addition to N_2O, other nitrogen oxides NO_x are also emitted from ammonia and nitric acid plants. These emissions are estimated as 1-2 kg per tonne of ammonia and 6-9 kg per tonne of nitric acid.

Abatement techniques for reducing these emissions are available and fertilizer plants need to take necessary modifications to implement them. New technologies are under development to control N_2O emissions from this process. A number of factories have benefited from carbon trading mechanisms in order to abate their N_2O emissions, and others are expected to follow.

PHOSPHORUS FERTILIZER PLANTS

There are two types of pollutants from the phosphate fertilizer plants: gaseous sulphur dioxide (SO_2) from sulphuric acid plants and fluorine from rock phosphate.

Gaseous SO_2 emitted to the atmosphere contributes to acid rain which causes acidity in water bodies and death of fish and other aquatic animals. However, acid rain provides sulphur to the soil and supplements soil fertility.

Rock phosphate contains 3-4.5% fluorine which is released on acidulation during manufacture of wet process phosphoric acid and single or triple superphosphate. About 4-5 tonnes of phosphogypsum are produced from each tonne of P_2O_5 in phosphoric acid.

Disposal of phosphogypsum is a problem due to its fluorine content. During manufacture of single or triple superphosphate, the fluorine gas escapes to atmosphere.

POTASH FERTILIZER PLANTS

Unlike nitrogen and phosphate fertilizers, potash production has no known pollution potential on the environmental quality.

ENVIRONMENTAL PROTECTION STANDARDS

The environmental protection standards are essential to ensure that the pollutants discharged from the industry into the environment are within the capacity of the environment to assimilate them through natural purification processes. In developed countries and many developing countries, environmental protection legislations have been enacted to enforce environmental protection standards. In India, the first legislation came into existence in 1974, specifically to protect water resources from industrial pollution. This legislation is referred to as the Water (Pollution Prevention and Control) Act, 1974. This was followed in 1981 by the Air (Pollution Prevention and Control) Act. In 1986, the Environment (Protection) Act was enacted, which is an umbrella Act covering all facets of the environment.

Water (Pollution Prevention and Control) Act, 1974 is designed for prevention and control of water pollution and maintaining or restoring the wholesomeness of water. Under this Act, the Central Pollution Control Board (CPCB), a statutory body, was set up at the national level. Subsequently, State Pollution Control Boards (SPCB) were established at state levels. This act provides that the CPCB in consultation with the State Governments is empowered to lay down, modify the national standards for streams, sewers and land. The CPCB has laid down Minimal National Standards (MINAS) for major sectors of the industry including fertilizers. The SPCBs follow these standards or modify them to make them more stringent, if the situation so warrants.

Air (Pollution Prevention and Control) Act, 1981 provides for the prevention, control and abatement of air pollution and for preservation of air quality. Under the provisions of this Act, the CPCB shall lay down the standards for the quality of air and devise means to prevent, control and abate pollution without any need to consult the States. State Pollution Control Boards shall lay down the standards for emission of air pollutants in consultation with CPCB and will act under the direction of CPCB.

The Environment (Protection) Act, 1986 is an umbrella Act and provides for the protection and improvement of the environment and prevention of hazards to human beings, other living creatures, plants and property. This Act empowers the Central Government to prescribe the environmental protection standards that are statutory and applicable throughout the country. Environmental protection standards are thus legal and override all standards specified under other Acts. The Ministry of Environment and Forests is authorized to lay down the standards and enforce them. Penal provisions have been made for contravention of the Environment Protection Act, while some fiscal incentives and tax benefits are provided to encourage the industry to control environmental pollution.

As per Environment Protection Act, standards for nitrogen and phosphate fertilizer factories are shown in Tables 16.1 and 16.2.

Under the Air (Pollution Prevention and Control) Act, 1981 national ambient air quality standards for different categories of areas have been prescribed with respect to gaseous and solid pollutants: sulphur dioxide (SO_2), carbon monoxide (CO), oxides of nitrogen (NO_x), lead and suspended particulate matter (Table 16.3).

Table 16.1 Standards for nitrogen fertilizers (as per Environment Protection Act, 1986)

Parameter	*Concentration in the effluent not to exceed (in mg/L except for pH)*	
	Plants commissioned Jan.1, 1982 onwards	*Plants commissioned prior to Jan.1, 1982*
Liquid effluents		
pH	6.5–8.0	6.5-8.0
Total Kjeldahl N	100	150
Ammoniacal-N	50	75
Free ammoniacal-N	4	4
Nitrate-N	10$	10$
Cyanide (CN)	0.2	0.2
Vanadium (V)	0.2	0.2
Arsenic (As)	0.2	0.2
Suspended solids	100	100
Oil and grease	10	10
Chromium (Cr^{6+})	0.1	0.1
Total chromium	2.0	2.0
Gaseous emissions		
Parameter	*Prilling tower commissioned after 1-1-1982*	*Prilling tower commissioned prior to 1-1-1982*
Urea (particulate matter emissions)	50 mg m^3	150 mg per normal m^3
	0.5 kg per tonne of urea	2.0 kg per tonne of urea

$ 20 mg/L for CAN and ammonium nitrate fertilizers

Table 16.2 Standards for phosphate and complex fertilizers (as per Environment Protection Act, 1986)

Parameter	*Concentration in the effluent not to exceed, (in mg/L except for pH)*	
	Plants commissioned Jan.1, 1982 onwards	*Plants commissioned prior to Jan.1, 1982*
Liquid effluents		
(a) Straight phosphate fertilizers		
pH	7.0–9.0	7.0–9.0
Phosphate (as P)	5.0	5.0
Oil and grease	10	10
Suspended solids	100	100
Fluoride (as F)	10	10
Chromium (Cr^{6+})	0.1	0.1
Total chromium	2.0	2.0

(b) Complex fertilizers		
pH	6.5–8.0	6.5–8.0
Total Kjeldahl N	100	150
Ammoniacal-N	50	75
Free ammoniacal-N	4	4
Nitrate-N	10[$]	10[$]
Cyanide (CN)	0.2	0.2
Vanadium (V)	0.2	0.2
Arsenic (As)	0.2	0.2
Phosphate (as P)	5	5
Suspended solids	100	100
Oil and grease	10	10
Fluoride (as F)	10	10
Chromium (Cr^{6+})	0.1	0.1
Total chromium	2.0	2.0
Gaseous emissions		
(a) Phosphoric acid unit	Fluoride particulate matter	25 mg per normal m^3 150 mg per normal m^3
(b) Nitric acid unit	Nitrogen oxides	3 kg per tonne weak acid
(c) Sulphuric acid unit	Sulphur dioxide Acid mist	4.0 kg per tonne of 100% acid 50 kg per normal m^3

[$] 20 mg/L for complex fertilizers including CAN, ammonium nitrate and ammonium nitrophosphate fertilizer

ISO 14000 (Environmental Management Systems)

The International Standards Organization (ISO) is a non-governmental organization located in Geneva, Switzerland, which was established in 1947. The organization mainly functions to develop voluntary technical standards that aim at making the development, manufacture and supply of goods and services more efficient, safe and clean.

ISO has established ISO 14000 Environment Management System which is a structured approach to addressing the environmental bottom line. ISO 14000 refers to a family of voluntary standards and guidance documents to help organizations address environmental issues. Included in the family are standards for Environmental Management Systems, environmental and EMS auditing, environmental labelling, performance evaluation and life cycle assessment.

The establishment of ISO 14000 Environment Management System is an essential requirement for doing business in Europe and other regions of the world. ISO 14001 is the world's most recognized environment management system framework. India has adopted ISO Standards 14001, 14004, 14010, 14011 and 14012 issued by the ISO.

Table 16.3 National ambient air quality standards

Pollutant	Time (Weighted average)	Concentration in ambient air		
		Industrial area	*Residential, rural and other areas*	*Sensitive area*
Sulphur dioxide (SO_2)	Annual*	80 μg/m³	60 μg/m³	15 μg/m³
	24 hours**	120 μg/m³	80 μg/m³	30 μg/m³
Oxides of nitrogen (as NO_2)	Annual*	80 μg/m³	60 μg/m³	15 μg/m³
	24 hours**	120 μg/m³	80 μg/m³	30 μg/m³
Suspended particulate matter (SPM)	Annual*	360 μg/m³	140 μg/m³	70 μg/m³
	24 hours**	500 μg/m³	200 μg/m³	100 μg/m³
Respirable particulate matter (RPM) size <10 μm	Annual*	120 μg/m³	60 μg/m³	50 μg/m³
	24 hours**	150 μg/m³	100 μg/m³	75 μg/m³
Lead (Pb)	Annual*	1.0 μg/m³	0.75 μg/m³	0.50 μg/m³
	24 hours**	1.5 μg/m³	1.0 μg/m³	0.75 μg/m³
Carbon monoxide (CO)	Annual*	5.0 mg/m³	2.0 mg/m³	1.0 mg/m³
	one hour	10.0 mg/m³	4.0 mg/m³	2.0 mg/m³

* Annual arithmetic mean of minimum 104 measurements in a year taken twice a week 24 hourly at uniform interval.

** 24 hourly/8 hourly values should be met 98% of the time in a year. However, 2% of the time, it may exceed but not on two consecutive days.

ENVIRONMENTAL POLLUTION DUE TO FERTILIZER USE

Soil fertility decline is occurring over large parts of the world, particularly the developing world. It occurs mainly through intensive cultivation and inadequate application of nutrients to replace the amount removed in the harvests, and through deforestation and clearance of vegetation on sandy soils. Large amounts of soil nutrients are also lost to the terrestrial ecosystems through wind and water erosion. Low soil fertility is considered as one of the most important constraints on improved agricultural production. To sustain the future world population more fertilizers are required, which may become an environmental hazard, unless adequate technical and socio-economic measures are taken. It is estimated that, by the year 2020 at a global level, 70% of plant nutrients will have to come from fertilizers. Fertilizers are thus indispensable for sustained food production, but excessive use of mineral fertilizers has roused environmental concerns. Chief among these concerns are:

- Eutrophication (nutrient pollution that depletes water of oxygen and can kill aquatic life) of lakes and coastal waters by nitrogen and phosphorus run-off;
- Build-up of nitrogen that leads to acidification of soil (this can foster the release of elements like aluminium, which are toxic to plant roots and can lead to yield declines in crops such as sorghum);
- Increases in toxic chemicals in the soil through impurities left in fertilizer;
- Over-application of fertilizers can kill soil organisms, which harms the soil's ability to retain and recycle nutrients;

- Emission of nitrogen oxides into the air can contribute to global ozone depletion and greenhouse gases.

NITROGEN FERTILIZERS

When nitrogen fertilizers are applied to soil usually less than half of the applied nitrogen is used by the crop and the remaining amount is lost through run-off and leaching, and gaseous loss through ammonia volatilization and denitrification.

Nitrogen fertilizer loss through run-off is greater from slopy field not covered by vegetation than level fields covered with crop. The run-off nitrogen causes nutrient enrichment in lakes and steams, and finally coastal areas of the ocean. Nitrate form of nitrogen is prone to leaching loss and the loss is considerable from highly permeable soils (sandy loam or loamy sand). Nitrate lost through leaching pollutes the underground water. Fertilizer N has been held as main source of nitrate enrichment of groundwater, particularly in European countries and USA. The fertilizer N contributes largely towards nitrate enrichment of groundwater when the nitrogen application rates are high (200-700 kg/ha/yr) as in Europe, and heartland states of Iowa, Nebraska, Kansas and potato producing area of northern Maine in USA. In India, the highest rate of fertilizer N consumption is in Punjab where soils are light textured and highly permeable. Recently increase in groundwater nitrate concentration in well waters in Punjab has been reported.

In Europe, large areas of the North Sea coastlines and areas of the Mediterranean have suffered from eutrophication due to nitrates. In the USA, nitrates and phosphates are suspected of causing *Hypoxia* or the "Dead Zone" in the Gulf of Mexico. There is a great deal of controversy as to the cause, and even if these nutrients prove to be the cause, they may originate from several different sources apart from mineral fertilizers. Nutrient-enriched water, especially run-off from agriculture, is also incriminated in the *Pfiesteria* problem that killed a large number of fish in Chekaspeake Bay, USA, in the summer of 1997. It is highly unlikely that mineral fertilizers are primarily responsible for either of these problems.

There is concern that fertilizers are polluting both surface waters and water in aquifers, although the direct impact of the application of mineral fertilizers on the nitrate content of waters is poorly defined. World Health Organization (WHO) has recommended a limit of 50 mg of nitrate (NO_3) per litre of drinking water. In general, in developed countries, where mineral nitrogen fertilizer is suspected as a major source of water pollution, it is usually in areas of vegetable production or irrigated sandy soil, or where the optimum rates are exceeded. A distinction must be made between correct nitrogen fertilization and excessive animal excrement application. There is generally little danger of the nitrate pollution of groundwater due to the application of fertilizer on rain-fed crops in developing countries, because the application rates are well below the optimum. In irrigated agriculture, water management is an important issue for the loss of nutrients to the groundwater.

Ammonia volatilization is prominent when urea or ammoniacal N fertilizers are broadcast on calcareous soils and topdressed in flooded rice fields. Generally, more ammonia escapes to atmosphere from animal manure spreads than from the fertilizers.

Nitrous oxide has a greenhouse effect and is considered to be detrimental to the ozone layer. According to experts of the Intergovernmental Panel on Climate Change (IPCC), nitrous oxide is responsible for 7.5% of the calculated greenhouse effect caused by human activity. Its concentration in the atmosphere is increasing at a rate of about 0.2% per annum. Soils are the major global source of N_2O accounting for about 65% of all emissions. Nitrogen fertilizers can be a direct and indirect source of N_2O emissions; IPCC currently assume an N_2O emission factor of 1.25% of fertilizer N applied, but with a nine-fold range, from 0.25% to 2.25%. In general, fertilizer management strategies that increase the efficiency of N uptake by crops are likely to reduce emission of N_2O to the atmosphere.

Plants are one of our greatest assets in reducing greenhouse gases. As crops grow they absorb CO_2 (carbon dioxide) from the atmosphere to obtain the carbon they need to produce leaves, stems, roots and grain. Nutrients enhance this process by increasing crop growth and therefore the amount of atmospheric carbon absorbed and stored in plant material. When this plant material is worked into the soil it decomposes with a portion released back to the atmosphere and a portion used to build carbon-rich soil organic matter. This process not only improves our atmosphere but enhances the quality of our agricultural land by increasing soil organic matter. The use of nitrogen fertilizers therefore results in both the generation and the removal of greenhouse gases.

PHOSPHATE FERTILIZERS

The over-enrichment of surface waters leading to an excessive multiplication of algae and other undesirable aquatic plant species, with various undesirable consequences, is a phenomenon known as eutrophication. Whereas phosphate is generally the limiting nutrient in inland waters, nitrogen is generally the limiting nutrient in coastal waters.

Under temperate conditions, in inland fresh water, phosphate is usually the limiting nutrient and very low concentrations can cause problems of eutrophication. Surface water can be enriched with phosphate from both the point source (e.g. sewage treatment works) and diffuse sources (e.g. agricultural land). As the amount of phosphate from the point sources has declined in recent years, the percentage contribution from diffuse sources has increased. Since most soils fix phosphate strongly, only very small amounts of phosphate is likely to be lost from soil to cause environmental problems. Phosphate in the soil is rather immobile and the loss of water soluble phosphate through leaching is usually very small (less than 1 kg of P_2O_5 per hectare per year). Ignoring crop removal, the two primary pathways of loss of phosphorus from the soil are by erosion (wind and water) and in run-off. However, the excessive surface applications of animal manures can result in significant losses in particulate matter in run-off. Areas where intensive animal husbandry is practised may experience the addition of excessive amounts of phosphorus to the soil, usually in the form of heavy applications of animal waste, e.g. slurry or farmyard manure. Under these conditions, soils can have such a high content of phosphorus that losses may increase.

In tropical lakes, there is evidence that nitrogen can be the limiting nutrient. Phosphate concentrations in the water are often higher than in temperate regions while inputs of N from surrounding land may be low. Surface runoff (including soil erosion) from cropland, pasture and forest, can contribute to phosphate loading of surface waters. Best management practices are highly effective in eliminating this possibility and, at the same time, allow for the most efficient use of crop fertilization.

Rock phosphate contains a number of heavy metals such as arsenic, nickel, cadmium, lead, radium and aluminium as impurities (Table 16.4). Continuous application of rock phosphate or single superphosphate can result in accumulation of these heavy metals in soils. Crop plants take up these nutrients from soils and when consumed by humans and animals these are returned back to the soil through excrements.

Phosphorus fertilizers contain various impurities from the rock phosphate and acid used in manufacture. Cadmium (Cd) is of most concern. The levels of cadmium in fertilizer will vary with the rock phosphate source. As there is currently no commercial means of removing cadmium during fertilizer manufacture, the only control is to use fertilizer made from low cadmium rock phosphate, especially in areas of intensive use, e.g. vegetables.

Table 16.4 Average heavy metal concentration in rock phosphate deposits

Rock phosphate deposit	*Heavy metal concentration, mg/kg*						
	As	Cd	Cr	Pb	Hg	Ni	V
Russia (Kola)	1	0.1	13	3	0.01	2	100
United States	12	11	109	12	0.05	37	82
South Africa	6	0.2	1	35	0.06	35	3
Morocco	11	30	225	7	0.04	26	87
North Africa	15	60	105	6	0.05	33	300
Middle East	6	9	129	4	0.05	29	122
Average of 91% of RP reserves, mg/kg of P_2O_5	4.8	10.9	82.0	4.4	0.02	12.6	38.4
g/ha/yr applied with 20 kg P/ha	1	3.3	25	1	0.01	4	12
Tolerance limit in soil, mg/kg	–	2	100	100	2	50	50

Source: Kongshaug et al. (1992)

Table 16.5 Contents of heavy metals in common mineral fertilizers (ppm)

Fertilizer	Pb	Ni	Cu	Zn	Cr
Triple superphosphate			70		
Ammonium sulphate	23	44	1	2	5
Calcium ammonium nitrate	21	13	4	13	8
Calcium nitrate	35	60	12	1	26
Urea	18	8	1	40	4

Source: Hilal et al. (1996)

Use of phosphorus fertilizers may lead to a build-up of the heavy metal cadmium in soils. Cadmium is normally present in soils at levels of 0.1-1.0 mg/kg soil. Phosphorus fertilizers contain higher levels of cadmium as an impurity, so prolonged use over time increases soil cadmium levels. Cadmium input to soils also occurs from the atmosphere, e.g. near industrial centres, and in sewage sludge.

The use of phosphorus fertilizers may increase cadmium concentrations in farm produce, which in some situations may exceed maximum permissible levels. This is most noticeable with certain crops, e.g. potatoes and leafy vegetables (lettuce and spinach) and in the offal (kidneys and liver) of animals. Use of fertilizers with a low cadmium content, i.e. less than 100 mg of cadmium per kg of phosphorus (P) and preferably less than 50 mg of cadmium per kg of P is recommended, especially in vegetable production.

POTASH FERTILIZERS

Unlike nitrogen and phosphate, potash has no known deleterious effect on the quality of natural waters.

MINIMIZING ENVIRONMENTAL POLLUTION BY FERTILIZER USE

Best Management Practices for fertilizer use effectively reduce the risk of environmental pollution. These practices can be summarized in simple words: the *right product(s)* should be used at the *right rate, time and place*. This implies that manufactured fertilizers should be managed conjointly with all other sources of plant nutrients.

Right product(s): The fertilizer source and product should be matched to the crop need and soil properties. The nutrient interactions and balance of nitrogen, phosphorus, potassium and other nutrients according to soil analysis and crop needs should be taken into account. Balanced fertilization is one of the keys to increasing nutrient use efficiency. It is advisable to adapt the form of nitrogen (speed of uptake, susceptibility to leaching or volatilization, acidification potential, etc.) to the site and crop specificities.

Right rate: The amount of fertilizer applied should be matched to the crop needs. Too much fertilizer leads to leaching and other losses to the environment and too little results in lower yields and crop quality and less crop residue to protect and improve the soil structure. Realistic yield goals, soil testing, omission plots, crop nutrient budgets, tissue testing, plant analysis, applicator calibration, variable rate technology, crop scouting, record-keeping and nutrient management planning are elements of Best Management Practices (BMPs) that help determine the right rate of fertilizer to apply.

Right time: The nutrients should be made available when the crop needs them. Nutrients are used most efficiently when their availability is synchronized with crop demand. Application timing (preplant or split applications), controlled-release technologies, stabilizers and product choice are aspects of BMPs that influence the timing of nutrient availability.

Right place: The nutrients should be placed where crops can use them. An appropriate application method is critical for efficient fertilizer use. The cultivar, cropping system and soil properties dictate the most appropriate method of application, but incorporation is usually the best option to keep nutrients in place and increase their efficiency. Conservation tillage, buffer strips, cover crops and irrigation management are other measures that will help keep fertilizer nutrients where they were placed and accessible to growing crops.

The Future

Fertilizer research is currently focusing on reducing the harmful environmental impacts of fertilizer use and finding new, less expensive sources of fertilizers. Concerted efforts are being made to make fertilizers more environment-friendly, develop improved methods of application, supplying fertilizer in a form which is less susceptible to run-off, and making more concentrated mixtures. It is expected that the environment pollution potential of fertilizers will be greatly reduced in future.

Selected References

Ayoub, T.A. (1999). Fertilizers and the environment. *Nutrient Cycling in Agroecosystem* **55**:117-121.

Hodge, C.A. and Papovic, N.N. (1994). *Pollution Control in Fertilizer Production.* Marcel Dekker Inc., New York, pp. 502.

Isherwood, K. 1998. *Fertilizer Use and the Environment*, International Fertilizer Industry Association, Paris, pp. 51

UNIDO/IFDC (Eds) (1998). *Fertilizer Manual*, 3rd ed., Kluwer Academic Publishers, Dordrecht, the Netherlands.

UNEP 1998. Mineral Fertilizer Production and the Environment. Part 1.The Fertilizer Industry's Manufacturing Processes and Environmental Issues, Technical Report No. 26, UNEP-UNIDO-IFA.

Index

Acid forming fertilizers 167
Acids used for fertilizer production nitric acid 27-28
  phosphoric acid 29-32
  sulphuric acid 28-29
  superphosphoric acid 32-33
Air (Pollution Prevention and Control) Act (1981) 173
AM as nitrification inhibitor 61
Amide-N fertilizers 46
Ammonia, anhydrous
  characteristics 43
  direct application 43
  industrial uses 43-44
Ammonia as raw material 47
Ammonia converter 42
Ammonia manufacture
  ammonia synthesis 42
  gas purification 41-42
  shift conversion 41
  synthesis gas preparation 37-41
Ammonia production
  early attempts 35
  feedstocks 36
  Haber-Bosch process 36
Ammonia volatilization 177
Ammonium-N fertilizers 46
Ammoniation of superphosphate 64
Ammonium bisulphite 101
Ammonium chloride
  characteristics 57
  manufacture 56-57
Ammonium molybdate 114
Ammonium nitrate
  characteristics 54
  manufacture 53-54
Ammonium phosphates 68-70
Ammonium phosphate-sulphate
  characteristics 85
  manufacture 85
Ammonium phosphate-sulphur 100
Ammonium polyphosphate
  characteristics 70
  manufacture 70
Ammonium polysulphide 101
Ammonium sulphate
  characteristics 52
  manufacture 51-52
Ammonium sulphate as by-product from
  caprolactum 52
  coke ovens 51
Ammonium thiosulphate 101
Application of nutrients
  micronutrients 148
  nitrogen 146-147
  phosphorus 147
  potassium 147
  secondary nutrients 148

Arsenic (As) concentration in rock
  phosphate 179
Available phosphorus in fertilizer 63-64
  soil 164

B

Band placement 145-146
Basic slag
  characteristics 72
  manufacture 72
Biuret in urea 48
Borax 112
Boric acid 112
Boron fertilizers 112-113
Brine as potassium source 26
Broadcast application of fertilizers 144-145
Burned lime (*see* quick lime)

C

Cadmium (Cd)
  build up in soils 179
  concentration in rock phosphate 179
Calcitic lime (*see* limestone)
Calcination 21
Calcium chloride 114
Calcium fertilizers 103
Calcium ammonium nitrate
characteristics 55
manufacture 54-55
Calcium chloride 114
Calcium metaphosphate
  characteristics 71
  manufacture 71
Calcium nitrate
  characteristics 56
  manufacture 56
CDTA as chelate 107
Carbon dioxide removal 41
Carbon monoxide removal 41-42
Carnallite 25
Chelates in fertilizers (see synthetic chelates)
Chlorine fertilizers 113-114
Chlorophyll meter 155
Chromium (Cr) concentration in mineral
  fertilizers 179
  rock phosphate 179
Citrate soluble phosphate fertilizers 64
Clear liquid fertilizers 97
Coal as feedsock for ammonia 17
Coal gasification 40
Coal reserves in India 17
Coke from coal 40
Coke oven gas 16
Colemanite 112-113
Complex fertilizers 84-89
Concentrated superphosphate
  characteristics 68
  manufacture 68
Conditioners 92
Copper acetate 110
Copper ammonium phosphate
Copper chelates 110
Copper fertilizers 110-111
Copper nitrate 110
Copper oxide 110
Copper sulphate 110
Crotonyledene diurea (CDU) 61

D

Dicyandiamide 61
Denitrification 177
Diammonium phosphate (DAP)
  characteristics 69
  manufacture 69
Dolomite 103-104
Dolomitic lime (see dolomite)
DRIS system 154-155
DTPA as chelate 107
Dual-salt process 57
Dunnage 139

E

EDDHA as chelate 107
EDTA as chelate 107
Elemental sulphur
  as sulphur fertilizer 99-100
  characteristics 23
Energy requirement of different
  feedstocks for ammonia 37
Environmental management systems
  (ISO 14000) 175
Environmental pollution by
  nitrogen fertilizer plants 171-172
  phosphate fertilizer plants 172
Environmental pollution by fertilizer use
  nitrogen fertilizers 177-178

phosphate fertilizers 178-180
potash fertilizers 180
minimizing 180
Environment Protection Act (1986) 173
Environmental protection standards 173-174
Epsom salt 104
Eutrophication 176

F

Farmer's service centers 128
Feedstocks for ammonia production gaseous 15
liquid 16
solid 17-18
Ferric oxide 108
Ferric sulphate 108
Ferrous ammonium phosphate 108
Ferrous ammonium sulphate 108
Ferrous oxide 108
Ferrous sulphate 108
Fertigation 151-152
Fertilization (*see* fertilizer application)
Fertilizers and
agricultural development 11-12
food production 6-11
sustainable agriculture 12
Fertilizer application
amount to apply 152-155
methods of 145-147
time of 150-151
Fertilizer application based on
plant analysis 154-155
soil-tests 153
Fertilizer consumption trend in
India 7-11
the world 2-3
Fertilizer demand forecasting
need for forecasting 129
forecasting methods 130-132
Fertilizer efficiency improvement by integrated plant nutrient management (IPNM) 158-159
real time N management 155-157
remote sensing and precision agriculture 159
site-specific nutrient management (SSNM) 157-158
Fertilizer import in India 8
Fertilizer legislation
Fertilizer Control Order 116-117
standard specifications 117-123
Fertilizer logistics
movement 142
packaging 141
storage 137-140
Fertilizer marketing
marketing channels 127-128
prices 134-135
subsidy 135-136
Fertilizer production trend in India 8
Fertilizer promotion 133-134
Fertilizer retail network 128
Fertilizer sale points 128
Fertilizer quality control
by Government 125
by manufacturer 124-125
tolerance limits for nutrients 123
Fertilizers spare the land 13-14
Fertilizer use and environmental pollution 176-180
Fertilizer use, historical perspective 1-2
Fillers for mixed fertilizers 94
Fluoraptite in rock phosphate 21
Foliar fertilization 148-149
Food grains production in India 9
Food security 4-5
Fractional crystallization process 75-76
Frasch process 22-23
Frits as micronutrient carrier 107-108
Froth flotation process 77

G

Gas purification 41-42
Granulation 50
Greenhouse effect 177-178
Guanylurea 61
Gypsum as
calcium source 103
sulphur source 101
Gypsum process for ammonium sulphate production 52

H

Haber-Bosch process 36
Handling of fertilizers 142-143
Hargreaves process 81
Heavy fuel oil 16-17

Heavy metals concentration in
mineral fertilizers 179
rock phosphate 179
Heavy media separation 77
HEDTA as chelate 107
Hydrogen for ammonia synthesis 37
Hydrogen sulphide 23

I

Insoluble phosphate fertilizers 64
Integrated plant nutrient management (IPNM) 158-159
Iron chelates 108
Iron fertilizers 108-109
Iron pyrites
chemical composition 102
reserves in India 102
Isobutyledene diurea (IBDU) 61

K

Kainite 25
Kieserite 104

L

Langbeinite 25
Langbeinite process 80
Law of minimum (Liebig's) 1
Lead (Pb) concentration in
mineral fertilizers 179
rock phosphate 179
Leaf color chart (LCC) 155-157
Limestone 103
Liming materials 103
Limits of organic agriculture 12-13
Liquid nitrogen wash 42
Liquid N fertilizers 46
Long-term effects of fertilizer use on
available nutrients in soil 163-164
crop yield 162-163
soil acidity 165-168
soil erosion 169
soil organic matter 163-164
soil physical properties 168
soil structure 168
soil quality 169-170
Long-term fertilizer experiments in India 162
Rothamsted fertilizer experiments 161-162

M

Magnesia 104
Magnesium chelates 104
Magnesium chloride 104
Magnesium fertilizers 103-104
Magnesium nitrate 104
Magnesium oxide 104
Major nutrients 3
Manganese chelates 111
Manganese chloride 111
Manganese fertilizers 111-112
Manganese oxide 111
Manganese sulphate 111
Mannheim process 80
Merseberg process (*see* gypsum process)
Methanation 41
Micronutrients 3
Micronutrient fertilizers 105-115
Mineral nutrients requirement 3-4
Mercury (Hg) concentration in rock phosphate 179
Mixed fertilizers
advantages of 84
formula construction 94-95
guidelines for mixing 93
ingredients 91-92
Mixed fertilizers production
bulk blending 96
fluid mixed 97
granulated mixed 96
pulverized mixed 96
Molybdenum fertilizers 114
Molybdenum frits 114
Molybdenum trioxide 114
Monoammonium phosphate (MAP)
characteristics 69
manufacture 68-69
Muriate of potash
characteristics 79
manufacture 75-79
secondary processing 79

N

Naphtha 16
National ambient air quality standards 176
Natural gas 15-16
Natural organic complexes 107

Neem-cake coated urea 61
Neutral fertilizers 167
Nickel (Ni) concentration in
  mineral fertilizers 179
  rock phosphate 179
Niter 25
Nitrapyrin 61
Nitrate-N fertilizers 46
Nitrate pollution of ground water 177
Nitric acid
  characteristics 28
  manufacture 27-28
Nitrification 167
Nitrification inhibitor 61
Nitrophosphates
  characteristics 88
  manufacture 87-88
Nitrogen fertilizers
  acidity and basicity 167-168
  classification 45-46
Nitrogen solutions 58-59
  characteristics 59
  manufacture 59
Nitrous oxide as greenhouse gas 177
NPK complex fertilizers
  characteristics 89
  manufacture 89
N-Serve (*see* nitrapyrin)
NTA as chelate 107
Nutrient removal by crops 4

O

One-third to two-third method of sulphur
  recovery 23
Organic carbon in soil 163-164
Organic matter in soil, impact of fertilizer
  use 163-164
Orthophosphoric acid (*see* phosphoric acid)

P

Partial oxidation process 39
Phosphate fertilizers
  citrate solubility 63
  classification 63-64
  placement 147
  watersolubility 63
Phosphate rock (*see* rock phosphate)
Phosphoric acid
  characteristics 32
  raw materials 29
Phosphoric acid production
  furnace process 32
  wet process 30-32
Plant nutrients
  classification 3
  requirement by crops 3-4
Plant nutrient content
  nitrogen fertilizers 46
  phosphate fertilizers 64
  potash fertilizers 75
  sulphur fertilizers 99
Polyhalite 25
Polymer-coated urea 61
Polyphosphoric acid (*see*
  superposphoric acid)
Population growth in Asia, Africa,
  Europe and world 5
Potash extraction from brine 77-79
Potash fertilizers 74-83
Potash minerals 25
Potash resources of world
  brines 26
  salt deposits 24-25
Potash mining
  shaft mining 25
  solution mining 25
Potassium chloride (*see* muriate of potash)
Potassium magnesium sulphate (*see* sulphate of
  potash-magnesia)
Potassium nitrate
  characteristics 83
  manufacture 83
Potassium sulphate (*see also* sulphate of potash)
  characteristics 81-82
  manufacture 80-81
Potential acidity of commercial fertilizers 167
Preplant placement of fertilizers 145-146
Prills 50
Pyrites as sulphur source (*see* iron pyrites)

Q

Quick lime 103

R

Raw materials for nitrogen fertilizers 15-18
 phosphate fertilizers 18-21
 potash fertilizers 24-26
Real-time N management using
 chlorophyll meter 155
 leaf color chart (LCC) 156-157
Refinery gas 16
Remote sensing and precision agriculture 159
Rock phosphate
 beneficiation 20-21
 characteristics 21
 mining 20
 world reserve 18-19
Rock phosphate for direct application 72-73

S

Secondary nutrients 3
Secondary nutrient fertilizers
Seed band placement 146
Shift conversion 41
Sicilian process 22
Side dressing 146
Single superphosphate
 characteristics 66-67
 manufacture 65-66
Site-specific nutrient management
 (SSNM) 157-158
Slaked lime 103
Slow release N fertilizers 60-61
 coated materials 61
 low solubility materials 60-61
Slurry fertilizers 97
Sodium
 chloride 114
 molybdate 114
 nitrate 58
 pentaborate 112
 tetraborate 112
Soil acidification
 process 165
 factors that accelerate 166
 role of fertilizers in 166-168
Soil fertility changes after long-term
 fertilizer use 164
Soil organic matter, impact of fertilizers 163-164
Soil quality
 indicators 167
 long-term effect of fertilizers on 167
Soil-test based fertilizer recommendation 153
Solubor 112
Sour gas 23
Specifications for fertilizers as per FCO
 ammonium sulphate 117
 ammonium phosphate-sulphate 119
 bentonite-sulphur 123
 borax 121
 diammonium phosphate 119
 diammonium phosphate fortified with
 boron 122
 monoammonium phosphate 120
 neem-coated urea 122
 nitrophosphate 119
 NPK complex fertilizes 120-121
 NPK complex fertilizers fortified with
 boron 122
 NPK complex fertilizers fortified with zinc 121
 potassium chloride 118
 potassium sulphate 118
 single superphosphate 118
 urea 118
 urea-ammonium phosphate 119
 zinc sulphate 121
Stacking 139-140
Standard concentration of pollutants in the effluents
 of nitrogen fertilizer plants 174
 phosphate fertilizer plants 174
 complex fertilizer plants 175
Standard concentration of pollutants in the
 emissions from
 nitrogen fertilizer plants 174
 nitric acid units 175
 phosphoric acid units 175
 sulphuric acid units 175
Steam reforming process 37-38
Subsurface band placement 146
Sulphate of potash-magnesia (*see also* potassium
 sulphate)
 characteristics 81-82
 manufacture 80-81
Sulphur
 deposits in the world 22

mining 22
recovery from sour gas 23
Sulphur-bentonite 100
Sulphur-coated urea 61
Sulphur fertilizers classification 99
Sulphuric acid
characteristics 29
manufacture 29
Superphosphate (*see* single superphosphate and triple superphosphate)
Superphosphoric acid
characteristics 33-34
manufacture 33
Surface band placement 146
Suspension fertilizers 97
Sylvite 25
Sylvinite 25
Synthesis gas preparation processes 37-41
Synthetic chelates 106-107

T

Time of fertilizer application 150-151
Top dressing 146
Transport of fertilizers 142
Trona process 78
Triple superphosphate
characteristics 67
manufacture 67

U

Urea
Characteristics 50
manufacture 48-50
safe concentration for foliar spraying 149
Urea-ammonium phosphate
characteristics 86-87
manufacture 86
Ureaform 60
Urea hydrolysis 50
Urea-sulphur 100

V

Vanadium (V) concentration in rock phosphate 179

W

Warehouse management 140
Water electrolysis for hydrogen production 40
Water pollution by nitrogen fertilizers 177-178
phosphate fertilizers 178
Water (Pollution Prevention and Control) Act (1974) 173
Water-soluble phosphate fertilizers 64
Wheat grain yield in the long-term fertilizer experiment 163
World population projections (2010- 2050) 7
WTO implications for fertilizer marketing 132-133

Y

Yield response 163

Z

Zinc carbonate 110
Zinc chelates 110
Zinc concentration in fertilizers 179
Zinc fertilizers 109-110
Zinc
oxide 110
phosphate 110
sulphate 110